BRIGHT IDEAS

Inspirations for SCIENCE

Published by Scholastic
Publications Ltd,
Marlborough House,
Holly Walk,
Leamington Spa,
Warwickshire CV32 4LS

© 1991 Scholastic
Publications Ltd

Written by Winnie Wade and
Colin Hughes
Edited by Juliet Gladston
Sub-edited by Catherine
Baker
Designed by Lynne Joesbury
Series designed by Juanita
Puddifoot
Illustrated by Liam Bonney
and Chris Saunderson

Designed using Aldus
Pagemaker
Processed by Pages Bureau,
Leamington Spa
Printed by Ebenezer Baylis
& Son, Worcester

**British Library Cataloguing in
Publication Data**
Wade, Winnie
   Science - (Inspirations).
   1. Science
   I. Title II. Hughes, Colin
   III. Series
   500

   ISBN 0-590-76406-3

CONTENTS

INTRODUCTION

Science

What does science mean to you as a primary teacher? This is an important question which you need to consider in order to achieve a balanced approach to primary science in your classroom. Until recently in England and Wales, primary science has tended to be considered either as a dilution of secondary school science with its concentration on facts and understanding, or as an investigative activity with an emphasis on exploratory and other skills. The latter view has been strongly defended by many teachers, with the result that knowledge has been excluded or reduced to a small part of the science work in their classrooms. They make statements such as, 'Knowledge plays little part in science in my classroom. I am more interested in developing a wide variety of skills'.

We would all be in favour of the development of a wide variety of skills, but it is necessary to consider the importance of knowledge in the way children understand and apply science. Science involves knowledge, facts and understanding, and it has its own unique methods which can be used to find and verify new knowledge. These methods include the exploratory skills or process skills which have been emphasised, to different degrees, in primary science teaching and learning in the past. The knowledge of science, which helps to make sense ofexperience, has been treated in many different ways in primary schools. A balance between knowledge and skills has been achieved only in a minority of classrooms, despite the fact that the knowledge, skills and methods of scientific exploration should surely be introduced to all pupils in order to provide worthwhile experience.

BACKGROUND

Science in the National Curriculum (1989) sees a place within the framework of science for exploration skills and knowledge and understanding. The exploration skills fundamental to science as an investigative, enquiry-based discipline include:
• observing and interpreting observations;
• assessing similarities and differences;
• finding patterns and making predictions;
• identifying problems and raising questions;
• hypothesising;
• planning and carrying out investigations by experiment;
• fair testing and testing of hypotheses;

• collecting, recording and communicating findings in a number of appropriate forms;
• drawing inferences based on the findings obtained.

These are grouped together in the following way in *Science in the National Curriculum*, under Attainment Target 1: Exploration of Science.
'i. plan, hypothesise and predict
ii. design and carry out investigations
iii. interpret results and findings
iv. draw inferences
v. communicate exploratory tasks and experiments.'

The knowledge and understanding to be covered within the Science National Curriculum should include aspects of:
• The variety of life
• Processes of life
• Genetics and evolution
• Human influences on the Earth

• Types and uses of materials
• Earth and atmosphere
• Forces
• Electricity and magnetism
• The scientific aspects of information technology including microelectronics
• Energy
• Sound and music
• Light
• The Earth in space.

Children's attitudes are also important if they are to engage purposefully in scientific activity. The attitudes which need to be developed to provide scientific ways of thinking and working include:
• curiosity
• perseverance
• critical reflection
• open-mindedness
• appropriately valuing the suggestions of others
• sensitivity to the living and non-living environment
• willingness to tolerate uncertainty
• respect for evidence
• creativity and inventiveness
• co-operation with others.

Learning science

Children learn by interacting physically and mentally with their environment. They interact by using their senses which can be stimulated by such things as observing a woodlouse, rolling a toy down a slope, or gazing at the stars. Children attempt to understand these observations by referring back to any relevant prior knowledge they may already have. It is, therefore, the role of the teacher to present enlightening learning experiences at the appropriate time to stimulate observation, investigation and thinking.

However, children are not empty vessels. They bring to the classroom their own knowledge, which must be respected. Gaps in learning and misconceptions need to be challenged from time to time by the introduction of activities which will add to the ideas the children have already about woodlice, rolling objects, stars and so on. Their learning will depend upon a number of factors such as their level of understanding at the outset of the activity, the way they are encouraged to interact with the stimulus, and the questions which are posed to them as they continue their investigation. At times it will be necessary to encourage perseverance, at others to highlight the lack of rigour in a supposedly 'fair test'. Sometimes it will be prudent not to give too much assistance until the children have done sufficient thinking while, at other times, new aspects of knowledge may be offered to help children with the development of conceptual understanding.

Planning the teaching of science

Despite the introduction of the National Curriculum in England and Wales, many primary teachers will still wish to continue teaching science through an integrated approach. The activities included in this book have been set out so that busy teachers can easily select those appropriate for the particular topic under consideration. A specific age range has been suggested for each activity, but it will be important for you to interpret these in relation to the ability of individual children in your class. By careful selection, progression of learning can be achieved right through from age four to eleven.

What is in this book for you?

This book addresses the development of scientific skills together with knowledge, so that conceptual understanding may be enhanced. The intention is to encourage an enquiring, investigative approach to science in the classroom whether taught through themes, topics or other methods of curricular organisation. Each of the activities is accompanied by factual outlines of the basic scientific ideas, which you may need to help you approach with confidence the teaching and learning of science in the primary classroom. This book will provide vital support for teachers getting to grips with primary science.

Chapters 1 to 10 provide 122 activities which you may choose from, as appropriate, to develop your children's learning in primary science. The activities have been written for use in the normal classroom, and specialised equipment is only used where absolutely necessary. Each of the chapters begins with an introduction that places the activities in context and discusses some of the ideas and concepts to be developed.

Each activity includes a suggested approximate age range, but the children's ability and prior experience are more important than their age when selecting appropriate activities. Your knowledge of the children is of vital importance. The same is true of group size, though we do make suggestions for this too, based on our experience and the nature of the activity.

Each activity is accompanied by a list of the equipment and materials required. At times it may be necessary to substitute items, until your stock of resources is built up. A major part of each activity is a set of *What to do* guidelines. Examples of questions to ask and ways of recording the children's work are also given, and may be modified as appropriate to your needs. Another important feature is the science content of the activity. Many primary teachers have little or no science background and this information is intended to help teachers in this situation to develop confidence in dealing with science. Finally, many of the activities include ideas for further work, and hints on safety in the classroom.

Chapter 11 gives a list of topics rich in science, with associated topic webs. This chapter indicates which activities could be appropriately used in each topic. A grid showing how the activities in the book relate to the National Curriculum Attainment Targets is also provided, to assist in curriculum planning.

Advice in the difficult areas of assessment and record keeping follows in Chapter 12, with suggestions for ways of assessing children in the classroom, and some useful examples of record sheets. Some helpful addresses of equipment suppliers and educational organisations are listed in Chapter 13 on Resources. Chapter 13 also contains a list of useful books and references which will provide additional support for the scientific activities.

Finally, there is a convenient set of photocopiable sheets at the end of the book. These relate directly to some of the activities and will save much time and energy in the busy primary classroom.

As a whole, this book is designed to help you achieve a balanced approach to primary science in your classroom. Remember, boiling an egg, washing your hands, mending the car and putting a man on the moon all require an understanding of science, an understanding which is based firmly in the primary school.

CHAPTER 1

The senses

The topic of the senses provides an excellent opportunity to introduce a spectrum of science activities into the classroom. The primary school years are an ideal time for children to learn about themselves and to appreciate the way their senses interpret the world around them. The five senses – sight, touch, smell, taste and hearing – enable us to react to changes in our surroundings and respond to outside stimuli such as changes in pressure or light intensity. You can help the children to explore the world of their senses in many ways, whether through tasting different flavoured crisps, testing their eyes, or having fun listening to different sounds. These activities will help children to find out about themselves and develop their ideas about how they use their senses.

ACTIVITIES

1. Smelly things

Age range
Five to seven.
Group size
Small groups.
What you need
A blindfold, six yoghurt pots (covered) containing 'smelly' things like curry powder, tea, coffee, onion, cinnamon or lemon juice.
What to do
Children have fun finding out who has the best sense of smell in the class. Blindfold one child and take the lid off one of the pots. See whether he can guess what is in the pot. Repeat this with the other pots. Which smell does he like the best?
Science content
Smells are detected by sensory hairs at the top of the nasal cavity. Molecules of a strong-smelling substance dissolve in the moist nasal cavity. This stimulates nerve endings to send nerve impulses to the brain, which then identifies the smell.
Further activity
The children could make a list of their favourite smells.
Safety
Children should be given guidance to ensure that they only smell harmless substances.
AT3/3a

2. Test your taste buds

Age range
Five to seven.
Group size
Pairs.
What you need
Straws cut up into small sections, solutions of sugary water, salty water, lemon juice, bitter cold coffee.

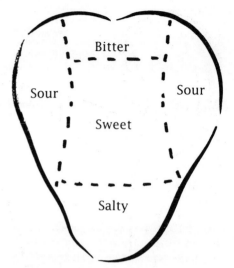

What to do
Ask the children to use separate pieces of the straw to put tiny drops of each liquid, in turn, on to their partners' tongue. They should aim for the front, back and sides of the tongues. (The children will need to rinse their mouths out between tastes.) Where can they taste each liquid best?
Science content
When food touches the taste buds which are located on the tongue, nerve impulses are sent to the brain which identifies the different tastes. Four basic tastes can be distinguished – sour, bitter, salty and sweet – but a substance must be partly dissolved in water to be detected as taste. Different areas of the tongue detect each taste.
Safety
During all tasting activities, great care should be taken to ensure hygiene and avoid activating allergies. *Only harmless substances should be tasted.*
AT3/3a

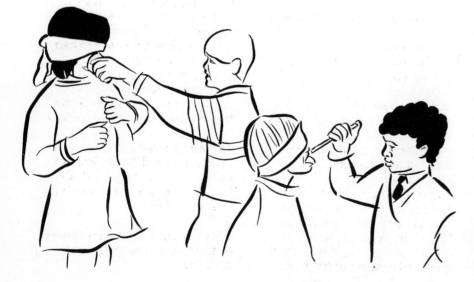

3. Hold your nose

Age range
Five to seven.

Group size
Pairs.

What you need
Different flavoured fruit gums cut up, a blindfold.

What to do
Ask one child in each pair to put a blindfold on and pinch her nose with her fingers. Her partner should place a sweet on her tongue. What flavour is the sweet? What happens if she does not pinch her nose? Is it easier to recognise the flavour?

Science content
The nerve endings detecting taste and smell are separate, but the overall enjoyment (or otherwise) of food depends on a combination of taste and smell. If the nose is blocked, most or all of the sense of smell is lost and we can only taste general sensations of sour, bitter, salt and sweet (see Activity 2, page 12).

Further activity
Try a similar activity with different flavoured crisps.

Safety
Ensure hygienic conditions during any tasting activity. Ask beforehand whether any child has a food allergy to colour additives or other substances.
AT3/3a

4. Test your feelers

Age range
Five to seven.

Group size
Pairs.

What you need
A fabric 'feely bag' containing an assortment of items such as a LEGO brick, coin, leaf, eraser, wax crayon, balloon.

What to do
Ask one child in each pair to put his hand in the feely bag and lift out one item at a time. However, *before* lifting it out of the feely bag, he should try to guess what the object is. Ask his partner to count how many he guessed correctly.

Science content
Our skin enables us to detect different sensations – heat, cold, pain, pressure, touch. There are touch receptor cells all over the skin, but some areas of skin have more than others, and so are more sensitive. For example, the touch receptor cells are grouped more closely on the finger-tips than on the backs of the hands.

Further activity
Children could try feeling an assortment of items with their feet to compare the sensitivity of feet and hands.

Safety
There should not be any sharp or pointed items in the bag.
AT3/3a

5. Skin senses

Age range
Five to seven.
Group size
Small groups.
What you need
Corks, pins, a blindfold.

What to do
Ask the children to press two pins into a cork, with the heads a small distance apart. They can then blindfold their friends and touch them *gently* with the pin heads in different places – on the arm, palm, fingertip, back of the neck etc. Let them try with one pinhead. How many pinheads can their friends feel? Are all parts of the skin equally sensitive?

Science content
See Activity 4, page 13.
Safety
Make sure the children are very gentle when touching each other with the pinheads. Supervise them carefully when sticking the pins in the corks.
AT3/3a

6. In focus

Age range
Seven to nine.
Group size
Small groups.
What you need
Metre ruler, photocopiable pages 152 and 153.
What to do
Pin the eye test chart to a wall and ask the children to stand about ten metres from the chart. Measure the distance to the chart. How many letters can they read? If they come one metre closer to the chart, can they read any more letters?
Science content
There is a lens in the eye which focuses images on to the sensitive area in the eye called the retina. Some people need spectacles because the lens is not working properly.
Further activity
The children could design their own eye test charts, perhaps with simple signs or pictures instead of letters.
AT3/3a

7. Are two eyes better than one?

Age range
Seven to nine.
Group size
Individuals.
What you need
Two coloured pencils.
What to do
The children should hold a pencil upright in each hand and stretch their arms out sideways at shoulder height. Ask them to move their arms together slowly, looking straight ahead, until they can just see the pencils. Let them

try again with one eye closed. Does this make any difference?

Science content
Humans have their eyes set at the front of the head, and each eye gives a slightly different view of the same object. This results in stereoscopic (three-dimensional) vision and enables us to judge distances. Your field of vision is the area you can see without turning your head.
AT3/3a

8. Guess the sound

Age range
Five to seven.
Group size
Small groups.
What you need
Eight tins, each containing a different object.
What to do
Can the children guess what is inside each tin by shaking it? How many of their guesses were right?

Science content
Sounds are made when things vibrate, sending sound waves through the air, which pass down the ear canal to the ear-drum and make the ear-drum vibrate. These vibrations pass into the inner ear, where nerve endings are stimulated and impulses go down the auditory nerve to the brain. Here they are translated into sounds.
Further activity
Let the children test their ears by covering one ear at a time. Do they hear better with one ear than the other?
AT14/1a

9. Where is the sound coming from?

Age range
Six to nine.
Group size
Pairs.
What you need
A blindfold, a small tin containing some beads.
What to do
Ask one child in each pair to wear a blindfold and look

straight ahead while her partner shakes the tin at different places around her (not too close!). Ask the blindfolded child to point to the direction from which she thinks the sound is coming each time.
Science content
Our ears help us to locate the direction of sounds. Sound waves will be detected slightly sooner by the ear nearer to the source than by the one that is further away. The brain analyses these differences and calculates the direction from which a sound comes. It may therefore be harder to detect the source of sounds coming from directly in front of or behind one.
Further activity
Make a 'sound journey' around the school. Tape different sounds and see if the children can guess the route taken on the sound journey.

The children could make a sound 'quiz' from a tape of different sounds. How many sounds can they collect? They could test the tape out on other groups of children.
AT14/2a

CHAPTER 2

My body and healthy living

Most children, and adults for that matter, are fascinated to investigate and learn more about their bodies. This interest has been heightened by the increasing number of people who have started to jog and keep fit, to question the use of additives in food, to select foods which are low in fats or sugars and to move away from traditional foods.

Younger children should be finding out about themselves, developing their ideas about how they grow, feed, move and use their senses and their knowledge about the stages of human development. They should be aware of how they keep healthy through exercise and by working and playing safely.

Older children should be introduced to the functions of the major organ systems and to simple ideas about the processes of breathing, circulation, growth and reproduction. They should realise that while all medicines are drugs, not all drugs are medicines. They should be aware of the terrible effects of drug abuse. Children should be introduced to the idea that all living things are made up of cells (or a cell), and that different cells carry out different roles in the organism.

ACTIVITIES

1. Parts of the body

Age range
Five to six.
Group size
Pairs or small groups.
What you need
Photocopiable page 154, scissors, adhesive, paper or card.
What to do
Ask the children to use photocopiable page 154 to cut out the external parts of the body and reassemble them on paper or card. Depending on age and ability, the children may name the parts of the body by pointing to them or by using the labels given.
Science content
The children will gain practice in naming the main external parts of the body including the feet, toes, hands, fingers, legs, arms, body, neck and head (incorporating the hair, eyes, eyebrows, ears, mouth, cheeks, nose).

Further activity
This piece of work could start with pictures or photographs of different people. Individual differences and similarities could be discussed, such as eye and skin colour, height, length of fingers and so on. Suitable songs and games may also be introduced, including the following:

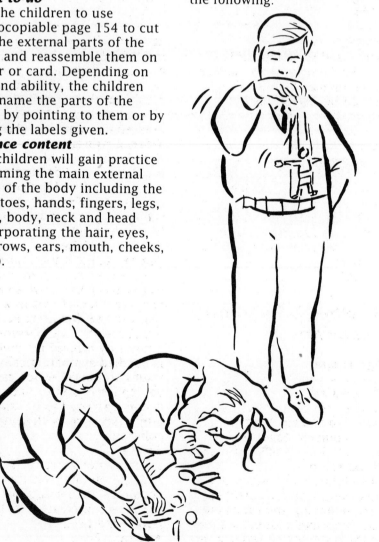

• Your legbone's connected to your...;
• Heads and shoulders, knees and toes...;
• Simon says....
Safety
Supervise the use of scissors, which should always be round-ended.
AT3/1a
AT4/1a, 2a

2. Living things reproduce their own kind

Age range
Five to seven.
Group size
Pairs or small groups.
What you need
Photocopiable page 155, scissors.
What to do
Ask the children to cut out the pictures of adult animals and their young from photocopiable page 155. They can then help the young to find their parents. The pictures could be placed under two columns or headings: 'Adults/parents' and 'Their babies/children'.
Science content
Living things reproduce their own kind. Adult humans produce humans as their offspring, lions produce lion cubs, and spiders produce small spiders. Mice cannot reproduce to give cats, and tigers cannot give birth to goldfish. However, children may get confused because they may know that a horse and a donkey reproduce to create an

ass, while two dogs of different breeds produce a cross-breed or mongrel. The general rule still applies, in that dogs reproduce to give dogs, and horses and their close relatives reproduce to form horse-like creatures. This happens because they are of the same species. Offspring which do not resemble their parents particularly closely, for example, goslings, may also lead to confusion. Children may be further confused by hearing that frogs produce tadpoles, unless they see the complete life cycle.

Further activity
Children's books and calendars are a good source of pictures, and slides or videos can also be useful. Classroom animals such as stick-insects, fish or gerbils may be effectively used to illustrate that living things reproduce their own kind. However, in the case of stick-insects, the eggs take several months to hatch.

Safety
If animals are to be kept in school, they should be purchased from reputable pet shops who offer sound advice on maintenance. Animals should be cleaned, fed and watered regularly and kept out of direct draughts. Children should wash their hands after handling or cleaning the pets. The school should have a clear policy to deal with grazes, bites etc.

The children should be encouraged to use scissors carefully when cutting out the pictures.
AT3/2a

3. What I did today/ yesterday

Age range
Five to seven.
Group size
Pairs or small groups.
What you need
Photocopiable page 156, scissors, large sheets of paper.
What to do
The children should cut out the pictures on photocopiable page 156 (removing any which are inappropriate) and place them in the order which reflects the pattern of their day. They can then stick them on a large piece of paper in the correct order.

Let the children take it in turns to tell one another what they did the previous day, highlighting the major events, such as, 'I got up; I ate my breakfast; then I washed; then I cleaned my teeth; then I dressed and went to school'.

Science content
The children will be considering and discussing a simple account of the pattern of their day. The correct sequencing of activities should be emphasised.

Further activity
Emphasise the importance of adequate sleep, regular meals, hygiene, regular exercise and appropriate pastimes as a follow-up to this exercise. It is important that the information innocently offered by children is handled sympathetically. If issues are raised which cause you concern, appropriate action may need to be taken, after discussion with senior colleagues.

Safety
Careful listening to children and appropriate, sensitive action will help safeguard the rights and interests of both pupils and teacher.
AT3/2b, 2c

4. Take a deep breath

Age range
Six to eight.
Group size
Pairs or small groups.
What you need
Pencil, paper, tape measure, rubber tubing, large tank or sink, sweet jar.
What to do
Ask the children to count a friend's rate of breathing when he is sitting quietly, and then again after he has been involved in activity. They can record the results in a table like the one shown below.

The children can then observe and feel what happens when they breathe in and out. What happens to the ribs and chest?

Working in pairs, let the children take each other's chest measurement, and then take it again when breathing in deeply. The chest expansion may then be calculated and displayed in a table similar to that shown below.

Who has the biggest chest expansion? Do big people generally have a larger chest expansion than small people? Do good athletes have a larger chest expansion?

The children can consider whether the people with the largest chest expansion have more air in their lungs than those with a small chest expansion. To test this they should first fill with water a deep sink or container and a large jar. They should then invert the jar in the sink and place a tube in the neck of the jar, as shown in the illustration.

As the children breathe out through the tube (without taking air in through the nose), air is forced out of the container. By calibrating the container from the top (zero) to the bottom, the displaced water which is equivalent to the volume of air in the lungs

Name	Chest size at rest (cm)	Chest size after breathing in (cm)	Chest expansion (cm)

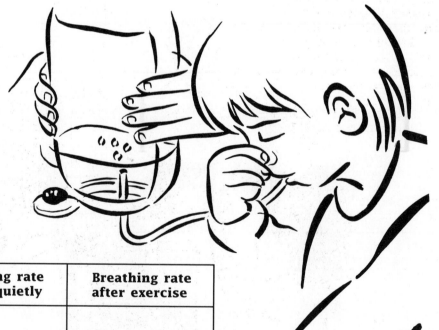

Name	Breathing rate sitting quietly	Breathing rate after exercise

Name	Chest expansion	Water removed = capacity of lungs

may be quickly read off. Comparisons may be made and related to the previous investigation, and the results recorded in a table like the one above.

Science content
When breathing in, the ribs and chest move upwards and outwards and a muscle called the diaphragm, between the thorax and the abdomen, moves downwards. This increases the volume of the thorax and lungs, and air enters due to atmospheric pressure. Air is expelled during breathing out because the ribs and chest move inwards and downwards and the diaphragm is raised due to the relaxation of the muscles. This decreases the volume of the thorax and lungs and increases the pressure; hence the air is automatically expelled.

The volume of air in the lungs is called the *lung capacity*. The lung capacity of an adult man is about 5,500cm^3, but during breathing when at rest only about 500cm^3 is taken in and expelled. A relationship might be seen between chest expansion and lung capacity and possibly also the ability to be successful at certain sports.

Further activity
Ask the children to look for patterns in the data collected.

Safety
The lung capacity investigation invariably ends up with a wet classroom, but it is great fun! Keep water away from electrical fittings, and make sure the children take care on slippery, wet floors. Check that children are well enough to carry out the required activity.

AT3/3a

5. The secrets of life

Age range
Seven to ten.

Group size
Pairs.

What you need
A classroom animal such as a stick-insect, fish, gerbil or guinea pig, videos of a variety of animals, and other secondary sources, for example, books, posters.

What to do
Ask the children to observe an animal closely. The activity could be focused by asking children to record any signs

that the animal is living. The children could make a note of the many things that it does over a certain time period.

Repeat this process, with another animal if possible, or use video material.

Alternatively, get the children to observe some of their classmates over a similar period. Are there any similarities with the animal? Are there any differences?

Science content
All living things breathe (respiration), feed, move, reproduce, get rid of waste products (excretion), and are aware of aspects of their environment (sensitivity/behaviour).

The children should see evidence of *movement, waste products, feeding* and *breathing* in many animals. The presence of babies would help to illustrate that living things *reproduce* and *grow*; how else would we or the gerbil be here? Examples of *behaviour* will be numerous but may be less obvious; put a 'new' object in the cage, shine a light (but not for long) or tap gently on the cage. Faecal material is of course a waste product, but is not considered as an excretory product as it is not formed in the body (the alimentary canal being a tube that runs *through* the body and not *in* the body). This is a difficult concept and need not be explained to children.

Further activity
Compare animals with plants. Children may know that the leaves of plants reproduce and grow. They could also be shown that plant leaves move towards the light. Plants of course make their own food, though insectivorous plants supplement their nutrient intake with the insects they capture. It will be less easy to convince children that plants take in air. Excretory or secretory products are small in quantity in plants, for example nectar or the crystals produced by some plants.

Safety
Children should be encouraged to respect and care for their animals at all times. Fast movements and excessive noise should be stopped, and the reasons for this explained. When investigating animals, the health of the animal should be paramount.
AT3/2a, 3a

6. Is my pulse rate the same as my heartbeat rate?

Age range
Nine to eleven.
Group size
Pairs.
What you need
Stopwatch or watch with a second hand, stethoscope.
What to do
Ask the children to:
• Predict whether the pulse

rate is the same as the heart rate.
• Explain the reason for this prediction.
• Predict/estimate the pulse rate or heartbeat rate.
• Measure (calculate) the pulse rate and heartbeat.
• Say how the test was made fair.
• Communicate the findings in an appropriate manner.
• State whether the prediction was correct or not.
• State any inaccuracies in the method used.

This activity is useful in encouraging a systematic, enquiring approach, and it helps to develop such exploration skills as prediction, accurate measurement, fair testing, recording results and making inferences. Children should be encouraged to pose questions such as, 'Is the heartbeat rate faster than the pulse rate?' They should then be encouraged to design an investigation to test their hypothesis.

Science content
The pulse rate and the heartbeat rate are the same if taken at the same time. As the heart beats, the pulse it creates may be detected at certain points on the body and the pulse rate determined. The pulse rate of an average person at rest is 75 pulses or beats per minute. The pulse is normally taken on the underside of the wrist, but many children find it difficult to detect there. The jugular artery pulse is easier to locate, on the neck under the jaw bone. A stethoscope may be used to count heartbeat and pulse rate.

The normal pulse rate will vary depending on fitness, metabolism and health. Some virus infections reduce the pulse rate to 50 per cent of its normal rate. The fitter the person, the quicker the pulse rate returns to normal after exercise or illness.

Further activity
Children could take their pulse rates before and after exercise and find out how long they take to return to normal. Individual differences could be considered, but it should be stressed that, if taken on another day, the results may be different. Children should be encouraged to take five readings and calculate the average. Sensitivity to children's lack of fitness is of paramount importance.

Introduce the idea of the circulatory system – heart, veins, arteries, capillaries.

Safety
Children with medical problems, those who have had a recent illness or who are known to be on medication should not take part in strenuous activity.

AT3/3a, 4a

7. Stages in the human life cycle

Age range
Seven to nine.
Group size
Small groups.
What you need
Photocopiable page 157, photographs, scissors, large sheet of paper.
What to do
Begin by focusing on an event such as a real birth or a death, or one in a picture, story or television programme, and encourage the children to think about and discuss the various stages in the human life cycle. They could start by thinking about their families, neighbours and friends. They might need to be helped to consider the stages they have been through, those their older sisters and brothers are a part of and the stages some of their relatives and neighbours are approaching. It may help to use resource material such as photographs. Ask the children to cut out the pictures on photocopiable page 157 which depict the stages in the life cycle of humans. They can then sequence the pictures in the correct order. They may also use written labels to accompany each stage.

Science content
The children will be introduced to the cycle: from parents to egg, to baby, to child, to teenager, and finally back to adult parents. The concept of a life *cycle* should be stressed. While children will probably be familiar (though vaguely) with the role of the egg in the life cycle of a chicken, they are less likely to be conversant with this stage in the human life cycle.

Further activity
The life cycle could be further developed by adding the further 'stages' of old age and death. Reference to other life cycles such as that of the frog, fish or chicken should precede the work and help to reinforce the idea that all living things have a life cycle. Wildlife films provide a source of excellent material, as many of these study the life of a particular animal through one full cycle. Further work could consider the span of the particular stages, and involve the children in creative thought on what it must be like to be a baby or a senior citizen.

NB The children's range of knowledge on reproduction will be variable and sensitivity should be exercised.
AT3/3b

8. Organs, organs everywhere

Age range
Nine to eleven.

Group size
Small groups.

What you need
Photocopiable pages 158 and 159, scissors, adhesive, very large sheets of paper.

What to do
The children can use the pre-prepared outline on photocopiable page 158 and stick on to it the other organs drawn on page 159. They can discuss among themselves where they think the organs are situated in the body and what they do, so that gradually the organs are put into place.

Once the children have completed this activity, one child can lie on a large sheet of paper while the others draw round her to produce an outline of the body.

They can then lay the outline on a table or pin it to a wall. With reference to the previous activity, the children should cut out organs of the appropriate size and place them on the body outline.

Science content
An organ is defined as part of an animal or plant which forms a structural and functional unit, such as the kidney, heart, leaf or root. Children should be introduced to organ systems and their associated organs as shown at the foot of the page.

Further activity
The internal nature of most of the organs of the body adds to the difficulties of learning. Reference to books, slides and film and video material will help to make the teaching and learning more concrete and supplement the activities described above. An obvious extension would be to explore the functions of the various organs and organ systems. This could be related to health and healthy living.

AT3/4a, 4b

Organ system	Organs
Sensory system	Ears, eyes, nose, tongue, skin
Nervous system	Brain, nerves
Respiratory system	Lungs
Digestive system	Foodpipe, stomach, intestines, liver
Reproductive system	Penis, testes, vagina, ovaries
Circulatory (blood) system	Heart and associated arteries and veins
Excretory system	Kidneys, bladder

9. A closer look

Age range
Ten to eleven.

Group size
Pairs.

What you need
Onion, blunt knife, microscope slide, cover slip, iodine solution, microscope.

What to do
Cut up the onion and put a piece of the 'skin' (which is found between the fleshy whorls) in a drop of water or iodine, and then place it on a microscope slide. Let the children view it under a microscope, initially using low power.

They can then draw or describe what is seen under the microscope at both low and high power.

Science content
The children should be introduced to the idea that all living things are made up of small units or 'building blocks' called *cells*. A cell is basically a small unit, as in a honeycomb, a small room or a cubicle in a prison.

Further activity
Children could look at a variety of cells using books, slides and bioviewers and find out their functions. Examples could include red blood cells which transport oxygen; white blood cells which help prevent disease; nerve cells, which carry messages; and muscle cells which bring about movement. This could tie in with earlier activities on the functions of body organs.

Safety
Care is required with glass cover slips and microscope slides, and also with iodine solution, which stains skin and clothes as well as cells! The removal of blood cells is not permitted because of the risk of infection, while the observation of cheek cells is discouraged or not permitted in some LEAs.

AT3/5a

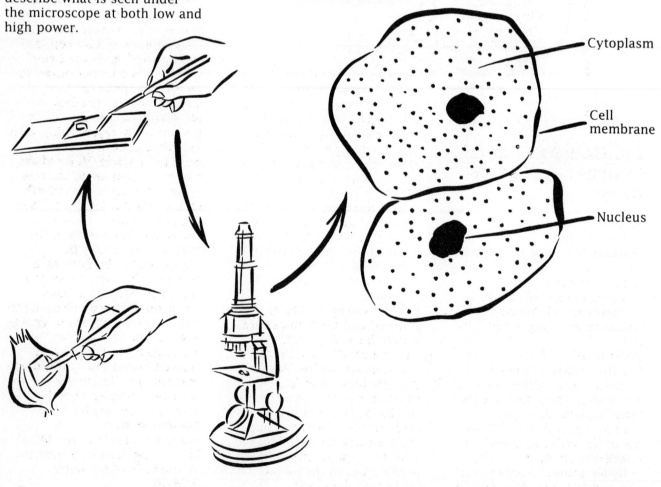

Cytoplasm

Cell membrane

Nucleus

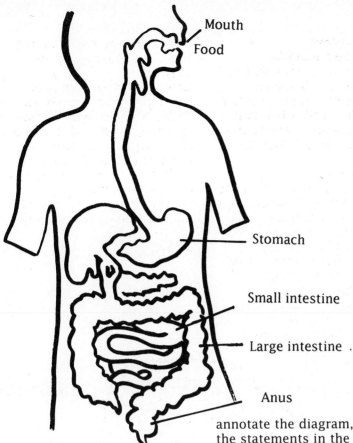

Mouth
Food
Stomach
Small intestine
Large intestine
Anus

10. How food gets to every cell of the body

Age range
Ten to eleven.
Group size
Pairs.
What you need
Bread, container (test tube or yoghurt carton), iodine, photocopiable pages 160 and 161.
What to do
Ask the children to place a piece of bread in their mouths and immediately describe its taste. Without biting or swallowing, they should mix the bread with saliva and notice any changes in taste.

Using photocopiable pages 160 and 161, the children can annotate the diagram, placing the statements in the correct sequence against the correct organs. This will enable them to understand the processes which food undergoes as it passes through the body. These are as follows:
• In the *mouth*, the food is bitten, chewed and broken up into small pieces by the teeth. Saliva is mixed with the food, and chemicals in the saliva begin to break it down further. These chemicals are called enzymes.
• In the *stomach*, food is churned and more juices are added, including acid.
• In the *small intestine*, more enzymes are added. Proteins, carbohydrates and fats are finally broken down into small, soluble substances which pass through the wall of the small intestine and into the blood.
• In the *large intestine*, much of the water in the food is taken into the blood.

• Waste food passes out of the body through the *anus*. This is brown in colour and it contains some moisture, but much of the goodness has been taken out of it.
Science content
Chemicals (enzymes) break down large complex molecules of food into simple, soluble chemicals which can be absorbed through the wall of the small intestine. This food is carried in solution in the blood to all parts of the body. The dissolved food passes through the capillary walls, which are one cell thick, and into the cells. Waste products pass the other way from the cells into the blood, where they are taken to the lungs (in the case of carbon dioxide), or the kidneys (in the case of urea).

A common misconception among children is the belief that food is only necessary to remove the physical and mental craving. The link between food and its role in providing energy, helping with growth and storing fat will need to be made. Many of the concepts involved in the role of food are abstract, and will require careful teaching. Use concrete examples where necessary. For example, the energy in food may be demonstrated by burning a peanut or biscuit so that the energy is released as heat. Unfortunately, while helping to reinforce the concept of energy in food, this can lead to the misconception that food is 'burned' in the body to provide that energy. Clearly, this is not the case; complex chemical reactions are involved.
Further activity
Examine the role of the blood in carrying dissolved food to all the cells of the body.
AT3/5c

CHAPTER 3

Materials, forces and structures

Children handle different materials every day in playing, at home and at school. These materials may be man-made, such as a plastic toy or a drinking glass, or they may be 'natural' materials such as wooden furniture or a woollen jumper. At an early age children enjoy building structures in their play. They see buildings, bridges and tunnels all around them in this modern world. They use forces every day in one way or another, and they see the effects of forces, such as the wind blowing the sails of a boat. It is for all these reasons that we should encourage children to explore and investigate what forces are, what they can do, what makes a structure strong and how materials are different from each other.

Children should collect a variety of everyday materials and look at the similarities and differences between them. They should experience different types of forces and investigate the strength of a structure.

Children should develop their knowledge and understanding of the properties of materials and the way these properties determine their uses. They should also increase their knowledge and understanding of forces, their nature, significance and effects on the movement of objects.

ACTIVITIES

1. Push or pull?

Age range
Five to seven.
Group size
Pairs.
What you need
Photocopiable page 162.
What to do
Explain to the children that a force is a push or a pull.
Ask them to look at the pictures on photocopiable page 162 and decide:
• Which picture shows a push?
• Which picture shows a pull?
They should write the word 'push' or 'pull' under the relevant pictures.
Science content
Pushes and pulls are forces. Forces are applied every day in one way or another. Lifting a weight, kicking a football or closing a door are all examples of the application of a force.
Further activity
The children could draw some pictures of their own showing 'pushes' and 'pulls'.
AT10/1a, 2a

2. Starting and stopping

Age range
Five to eleven.
Group size
Small groups.
What you need
A selection of toy cars, a piece of wood to use as a ramp, a metre ruler, blocks of wood of the same size.
What to do
The children can carry out a number of tests on each car.

• On a flat surface, the children can test how far each car will travel with a single push. The children should think about a 'fair test' and have the same child push each time. They can repeat this several times and record the results.
• Using the piece of wood as a ramp, the children can release each car from the top of the ramp *without* a push. They should measure the distance that the car runs from the end of the ramp, and repeat this several times, recording the results. If they change the slope of the ramp, what difference does this make to the distance travelled by the cars?
• Ask the children to put a block of wood at the end of the ramp. Does this stop the cars? How many blocks of wood are needed to stop the cars? (Relate this to road safety.)

Science content
• A force in the form of a push starts the car moving.
• A force is always needed to make things speed up or slow down.
• The greater the speed of the car, the greater the force needed to stop it.
• Speed tells you how far something travels in a given time.

Further activity
Other tests could be carried out by the children.
• Which is the fastest car? The children can measure with a stop-watch the time it takes for different cars to move a given distance, for instance from the top of the ramp to a fixed point on the floor. They can calculate the average speed of the cars using the equation:

$$\text{speed} = \frac{\text{distance travelled}}{\text{time taken}}$$

The children could investigate the surface material. They can design investigations to find out which surfaces have most or least friction. Ask them to cover the ramp or floor surface with different materials and repeat the above activities.
AT10/1a, 2a, 4b, 5c

3. What can forces do?

Age range
Five to seven.
Group size
Small groups.
What you need
Collection of toys such as a toy car, marbles, balloons, Plasticine, elastic band, small soft balls, a spring, photocopiable page 163.

What to do
Explain to the children that when they push or pull an object, they are using a force. Ask them to make a list of things in the classroom which move when pushed or pulled. They can then use some push and pull forces on the collection of objects to see what happens to each. They should fill in the correct box on the sheet provided.

Science content
Forces can:
• change the shape or size of an object, for example when blowing up a balloon or squeezing a piece of Plasticine;
• make things move faster or slower, for example when a car accelerates or a parachute slows down a fall;
• change the direction of something that is already moving, for example when hitting a moving tennis ball with a racquet.

Safety
Children should use a pump to blow up balloons. They should *not* use their mouths.
AT10/2a

31

4. Sink or swim?

Age range
Five to seven.
Group size
Small groups.
What you need
A large bowl of water, plastic sheet or newspaper to place under the bowl, a collection of objects made from different materials and different sizes, shapes and weights (pencil, coin, bottle top, paper-clip, cork, metal spoon), photocopiable page 164.
What to do
The children should guess which of the objects will float or sink. They can then put each object into the water one at a time and see which of them actually does float or sink. The children can then fill in photocopiable page 164.
Science content
Two factors, together, determine whether or not an object floats in a liquid:
• the density of the object;
• the shape of the object.

If the object is denser than the liquid, it will sink. If the object is less dense, it will float. Whenever an object is lowered into a fluid and displaces some of the fluid there is an upthrust caused by the pressure of the surrounding fluid. The upthrust is equal to the weight of the fluid displaced.

For a floating object, the upthrust equals the weight of the object. A ship is shaped so that it displaces a lot of water and there is a large upthrust as a result.
Further activity
Let the children carry out further investigations on floating and sinking, considering the effects of the shape and size of objects and the material of which they are made.
AT10/3b

5. Does it float in salt water?

Age range
Seven to nine.
Group size
Pairs.

What you need
A thin strip of balsa wood, drawing pin, large transparent plastic beaker and water, 30cm ruler, weighing scales, permanent felt-tipped pens (different colours), salt.
What to do
Ask the children to make a 'floater' by using a piece of balsa wood with a drawing pin to weight one end, making black felt-tipped pen marks at 0.5cm intervals on the wood. Next the children should fill the beaker about three quarters full using tap water. They should now put the floater into the water and mark on it the depth it floats at in the water using a different coloured pen. Now they can try the floater in water with different amounts of salt dissolved in it, using the same beaker and the same amount of water. First they can add 10g of salt, then 25g, 50g and 100g. They should record the depth of the floater each time.

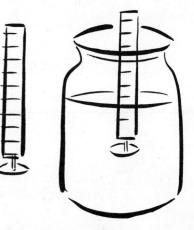

Science content
Salt water is denser than ordinary tap water, and gives a much stronger upthrust. The more concentrated the salt water, the higher the wood will float.
AT10/3b

6. Lifting it up

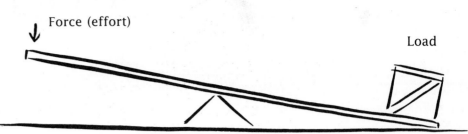

Force (effort)

Load

Pivot (fulcrum)

Age range
Five to eleven.
Group size
Small groups.
What you need
A collection of small tools and objects; for example, a tin with lid, spoon, spade from the sand-pit, digger from the sand-pit, pair of tweezers, ruler, heavy book, matchbox, pair of pliers, nutcracker.
What to do
Ask the children how work can be made easier. How can we get the lid off a tin or lift a heavy book? Using the tools and objects provided, the children should try them out and see how they can make work easier.

We can lift heavy things using a lever. To do this the children can rest the ruler on top of the matchbox and put the heavy book on one end of the ruler. If they press down on the other end of the ruler, can they lift the book?

Science content
A lever is a simple machine invented by early man to make work easier. Levers work by increasing the pushing force underneath a heavy weight so that it can be moved with a small effort. Levers have different forms, but all have a pivot on which the lever turns, a place at which the load is positioned, and a place where the force (effort) is applied. When you push down on one side, the other side goes up. Levers lift and move objects most easily when the pivot is close to the object or load and the pushing part is as far away as possible.

Further activity
The children could look for levers at home and at school; for example, the leverage used when pushing a door closed. The door is more difficult to close when you push with your fingers closer to the hinge.

Safety
Care must be taken when using any tools.

AT13/3c

7. Balance it out

Age range
Five to nine.

Group size
Small groups.

What you need
Ruler (30cm), a number of identical small weights, a pencil.

What to do
Ask the children to balance a ruler on a pencil so that the ruler acts as a see-saw. Where is the point of balance? Explain to them that this is called the pivot. Using a number of identical small weights, the children should put a weight at one end of the ruler and another weight on the other end of the ruler so that the ruler balances. If they move one of the weights nearer the pivot, what do they have to do to the other weight to make the ruler balance? If they put a number of weights on top of each other on one side of the ruler, what do they have to do to make the ruler balance again? Ask the children to put one weight on one end of the ruler. Where could they put two more weights to get the ruler to balance? They should measure the distances from the centre of the weights to the pivot.

Science content
The ruler acts as a see-saw or balance. A see-saw can be balanced even when one person weighs more than the other, by sitting them at different distances from the pivot. One weight placed at one end of the ruler balances two identical weights halfway between the pivot and the end of the ruler.

Further activity
The children could try altering the position of the pivot and adding weights at various positions on the ruler to balance it.
AT13/3c

8. Hooray and up it rises: hydraulics and pneumatics

Age range
Nine to eleven.

Group size
Pairs.

What you need
One 10 or 20cm^3 syringe, one small syringe (ideally 1 or 2cm^3, but a 5 or 10cm^3 syringe will do), small length of plastic tubing (3mm inside diameter) to fit tightly on to syringes, sink, water, object to lift (such as a light book).

What to do
Pull the plungers out of the two syringes and place them in a bowl of water along with the body of the syringes and the plastic tube. Connect the tube to the two syringes, ensuring a tight fit at both ends. With the apparatus still under water, push the appropriate plunger into the small syringe until it will go no further. Pour a small amount of water out of the other syringe and engage the plunger. The apparatus can now be taken out of the water and the children can gently press down the plunger of the larger syringe and watch what happens to the plunger of the smaller syringe. They can return the plungers to their original position by depressing the plunger of the smaller syringe. They can then repeat the activity, but this time they should put the plunger of the smaller syringe under a book. When they depress the plunger of the larger syringe, they will be able to watch the book rise!

Repeat the experiment, this time with air instead of water in the tube and syringes. Can the children see a difference? Which is more responsive?

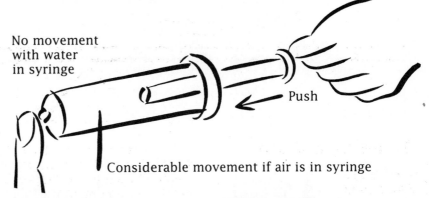

No movement with water in syringe

Push

Considerable movement if air is in syringe

Science content
The force exerted by the large plunger is transferred through the water to the small plunger where the distance of movement is exaggerated because of the relative diameters of the syringes.

Liquids cannot be depressed; pressure acts through them in all directions and they transmit pressure from one part of the liquid to another. Gases such as air may be compressed, and this explains why the movement of the plunger in the smaller syringe is delayed. A similar effect is produced if air gets into the hydraulic set-up. Hydraulic brakes, garage lifts and some lifting jacks make use of liquid pressure.

Further activity
Ask the children how they might apply the principle of hydraulics and/or pneumatics to their technology and model-making work. It could be used to make a lorry tip, or to act as a lift in an office or Santa's Grotto, or to make fairground rides go up and down.

Safety
With the pneumatic set-up, a strongly depressed plunger would propel the other plunger some distance and at some speed, and therefore care needs to be taken. The pneumatic set-up avoids the use of water and the inevitable spillage, but is not quite as efficient as the hydraulic model.

AT10/3a, 4a

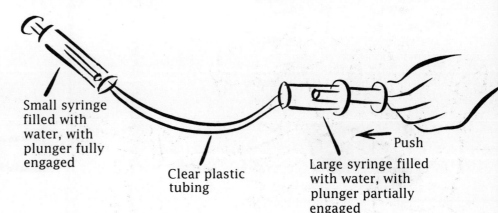

Small syringe filled with water, with plunger fully engaged

Clear plastic tubing

Push

Large syringe filled with water, with plunger partially engaged

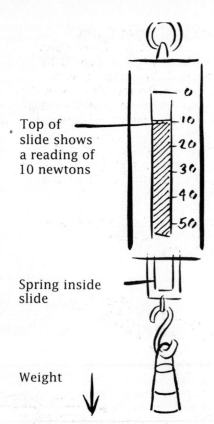

Top of slide shows a reading of 10 newtons

Spring inside slide

Weight

towards the centre of the earth. The force of gravity on a one kilogram mass is about ten newtons (10N). This measurement is referred to as the *weight* of the one kilogram mass. Weight is thus a force. A simple way of measuring a force is to use it to stretch a spring. The bigger the force, the more the spring is stretched. A spring with a scale attached to it and correctly calibrated is called a newton-meter. An apple weighs about one newton (1N); that is, one newton represents the force exerted on it by gravity. On the moon, where the gravitational pull is only about one sixth of that on Earth, you would weigh only one sixth of your normal 'Earth-weight'.
AT10/4d

9. Forceful weight

Age range
Nine to eleven.
Group size
Pairs.
What you need
A newton-meter, collection of various objects to weigh, for example, a 1kg mass, an apple, paper and pencil, Plasticine.
What to do
The children should hook each of the objects on to the newton-meter in turn and record the weight of each item in newtons by reading the scale on the newton-meter.
Science content
Forces are measured in units called newtons (N). A newton is the amount of force needed to give an object weighing one kilogram an acceleration of one metre per second, every second. Gravity is a force which pulls things down

10. Bridge that gap

Age range
Seven to eleven.
Group size
Pairs.
What you need
One sheet of standard A4 paper, scissors, adhesive, photocopiable page 165, container and weights, piece of polystyrene or stout card.
What to do
Ask the children to design and build the strongest bridge they can to cross a 15cm gap, using only *one* sheet of standard A4 paper, adhesive and scissors. They can test the bridge for strength by attaching a container to the bridge and loading it with weights. They should fill in the bridge testing sheet before and after testing.

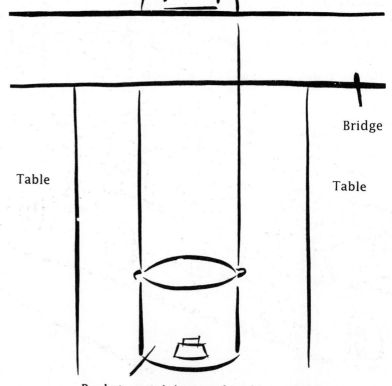

Piece of stout card or polystyrene

Bridge

Table

Table

Bucket containing sand and/or weights

Science content

Although there are several variations of each, there are four basic types of bridge: arch, beam, cantilever and suspension.

A bridge designer normally aims to provide, at the minimum cost, the structure which is best suited to the situation and the amount of traffic. Arch and beam bridges are ideal for short single spans or a series of short spans to overcome a wide but shallow river or valley. However, if the gap is deep as well as wide, a suspension bridge is an economic way of spanning it. The children will probably make a type of beam bridge.

A simple plank bridge bends because of the downward force of the weight of objects on the bridge. However, the plank will sag even when no one is standing on it. The force that causes this sag is the weight of the bridge itself. The weight of the bridge itself is called a 'dead load'. The weight of people, cars or trains that cross a bridge is a 'live load'. A bridge must be strong enough to support its own weight and it must be able to support, in addition, the greatest live load that might be placed on it. For example, during a traffic jam, a road bridge might be packed solidly with cars and lorries.

In a bridge, the force caused by the load pushes down on the roadway. This means that the bottom edge of a flat roadway is stretched while the top edge is compressed.

Further activity

Investigations could be carried out to see how folding a piece of paper increases its strength. Card could be folded in various ways and tested for load-bearing capacity. Conditions for the 'bridge' test could be changed; for example, with different distances between the bridge supports or using other materials such as balsa wood. Children could make cardboard models of girders and test their strength. Different shapes of arches could be made and tested too.

Safety

Make sure the children take care using scissors.

AT10/5b

11. A tower of strength

Age range
Seven to nine.
Group size
Pairs.
What you need
Eight sheets of A4 paper, scissors, golf ball, sticky tape, metre ruler or tape measure.
What to do
Ask the children to build the tallest tower they can, using eight sheets of scrap A4 paper. Their tower must stand up on its own and should hold a golf ball on the top for at least ten seconds. They should make a plan before they start building their tower. When they have built it, the children should measure the height of their tower. How could they improve on their design?
Science content
Children should be able to give an account of the strength of a structure. Paper can be rolled into tubes to make the tower. A cylindrical tube structure is strong because the sides of the tube cannot bend in two directions at the same time. The towers may also be strengthened with the addition of cross-pieces.
Further activity
Paper could be rolled into tubes and tested for strength in a number of ways – by changing the direction in which paper is rolled, or trying one tube inside another, or several tubes tied or stuck together.
Safety
Make sure the children take care when using scissors.
AT10/5b

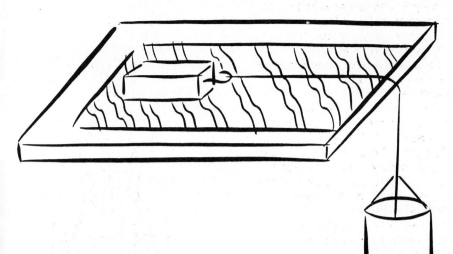

Friction is a force which tries to stop two objects moving over each other. Friction is very useful, for example, for the brakes and tyres of cars and bicycles. Friction can be reduced by lubricating the opposing surfaces with oil, as in a car engine. Where friction is reduced, the wheels of vehicles slip and spin on starting, as when it is icy.

Further activity

Let the children design investigations to find out whether wet conditions affect friction, and whether a larger 'pulling' or 'pushing' force is able to overcome friction.

AT10/5d

12. Rubbing together

Age range
Seven to nine.

Group size
Small groups.

What you need
Block of wood, a hook, string, plastic container, for example a margarine tub or yoghurt pot, a set of weights, different surfaces, for example carpet pieces, sandpaper, smooth board.

What to do
Ask the children to fix the hook on to the block of wood and tie one end of the string to the hook and the other end to the plastic container. Place each of the different surfaces in turn on top of a table, and see if the children can make the block of wood move over the different surfaces by hanging the plastic container over the edge of the table and putting different weights in the container. When do they need more weights to move the block? Which surface causes the most friction?

13. What is it made of ?

Age range
Five to seven.
Group size
Small groups.
What you need
Magazines, catalogues containing pictures of everyday objects, such as cars, clothes, toys, furniture etc, adhesive, scissors, paper.
What to do
Ask the children to cut out pictures of everyday objects from the magazines and catalogues. What is each object made of? Is it made from metal, plastic, cloth or wood? Are some objects made of more than one material? The children can stick pictures together in groups according to what they are made of.

Science content
We use a variety of materials in everyday objects. Some materials occur naturally and come from plants or animals, such as wood, cotton (plant), or wool, leather (animal), while others are found in the earth, such as stones. Many other materials are synthetic or man-made, like plastic and glass.
Further activity
Children could investigate different uses of materials.
AT6/2b, 3a, 3b

14. Metals and plastics

Age range
Seven to eleven.
Group size
Small groups.
What you need
A collection of small pieces of metal and plastic or small metal and plastic objects such as spoons, paper-clip, plastic straw, polystyrene (preferably the same sizes), a magnet, a nail, scissors, a bowl of water, photocopiable page 166.
What to do
Ask the children to carry out the following tests on each of the different samples of metals and plastics:
• What colour is the material?
• Is the material shiny or dull?
• Is the material smooth or rough?
• Does the material feel cold or warm?
• Can you bend the material without cracking or breaking it?
• Is the material strong? Can you tear it with your hands? Can you cut it with scissors?
• Does the material float or sink?
• Is the material picked up by the magnet?
• Can the material be easily scratched with a nail?
   The children should write down what they find on photocopiable page 166.
Science content
Most metals are found in the Earth's crust as ores (some exceptions are gold and platinum). Plastics are synthetic. Metals and plastics have many different properties, for example metals feel cold, plastics feel warm.
Further activity
Children could investigate the uses of metals and plastics in relation to their properties, for example, as saucepans, tool handles or jewellery.
Safety
Make sure that the children take care with scissors and be careful to ensure that there are no sharp edges on the metal or plastic objects.
AT6/1a, 2a, 3b, 4a, 4b

15. Spoons

Age range
Seven to nine.
Group size
Pairs.
What you need
Two clean teaspoons of similar shape and size, one metal and one plastic, a book, a pencil.
What to do
Ask the children to place each spoon on a book. They should then strike the spoons in the middle with a pencil. What sound do they hear? Which spoon makes a clanging sound? If they hold one spoon in each hand, does one feel heavier than the other? Can they see their reflection in either of the spoons?

Very gently, the children can try to bend each spoon without spoiling its shape. What do they think would happen if the plastic spoon was bent too much? Would the same happen if they bent the metal spoon too much?

Science content
The metal spoon should demonstrate several of the qualities of metals, including making a clanging sound when struck, reflecting light and bending without breaking. It should also be heavier than the plastic spoon and be a good conductor of heat.
Further activity
The children could stand the spoons in a beaker of warm water to see which of the handles heated up first.
AT6/4a, 4b

16. Fabrics and fibres

Age range
Seven to eleven.
Group size
Small groups.
What you need
Scraps of different types of fabrics (synthetic, wool and cotton), hand lens, paper and pencils, sticky tape, a simple microscope.
What to do
Let the children examine each fabric sample in turn. Ask them what it feels like. Is it coloured? Does it have a pattern on it? The children should look closely at the fabric with a hand lens. What can they see? Is it woven? Are there any spaces between the threads? Let them unpick a piece of the fabric so that they can look more closely at it under a microscope. See if they can draw a picture to show what the thread looks like through the microscope.

The children should then make a chart for each fabric sample, and stick a small piece of the fabric and its thread on to the chart, alongside the answers to their investigations. Do they know the name of the fabric? What is it used for?
Science content
Fibres are long, thin filaments that may be spun into yarn or thread. Threads can be woven into fabrics. Some fibres come from plants or animals and some are man-made. Wool comes from sheep and cotton comes from the cotton plant, whereas nylon and acrylic are man-made.
Further activity
Children could look at labels on clothes to identify natural fibres and man-made fibres.
AT6/2a, 2b, 3a

CHAPTER 4

Energy

The topic of energy is foremost in our minds as oil resources diminish, coal-fired power stations become less environmentally acceptable and critics of nuclear energy increase in number. While many young children are aware of these issues due to extensive media coverage, to them 'energy' means much more. It means being able to move around the football field, ice rink or discotheque, to manipulate their radio-controlled vehicle, to use the computer, to enjoy the flashing lights and to work the video recorder.

To some primary school teachers, a comprehensive coverage of the theme 'energy' will be new. It will involve you in new, exciting work which is certain to motivate all but the least interested child. Children will be introduced to the idea that energy from food is essential to every aspect of human life and activity. They will look at toys and devices which store energy in elastic bands, springs or balloons, so that their understanding of the nature of energy and its transfer and control is developed.

There is of course much more, for children at an early age will be investigating using magnets and making simple electrical circuits to light a bulb or ring a bell. They can see Rudolph's nose light up at Christmas, and help those explorers lost in the wood to escape by using a home-made compass. Children will be exploring the production of shadows and light reflecting from mirrors. You'll enjoy it too, that we guarantee; as long as you have the 'energy'!

ACTIVITIES

1. Food for thought, activity and life

Age range
Five to seven.
Group size
Whole class.
What you need
Pictures of different foods, of people eating, and of starving or malnourished people.
What to do
Discuss the children's favourite foods and ask them to draw or list their preferences. Collate the results for the class in the form of lists or drawings. Discuss the favourite foods of other nations and cultures.

Talk with the children about *when* they eat. Why is there a need to 'break one's fast'? Why shouldn't people eat late at night?

Discuss why people eat. It's more than just because they feel hungry; we need food to be active and strong, and to grow and stay healthy.
Science content
We need food to keep us strong, healthy and active.

Starving people do not have the energy to walk or run around. We also need food in order to grow.

We need a good variety of foods to stay healthy, including fresh fruit, vegetables and a protein source such as meat (remember that some people are vegetarians). We should not just eat crisps and hamburgers!
Further activity
Ask the children to save the wrappers from their favourite foods, and look at the contents listed on the wrappers.
AT13/1a, 1b
AT3/2b

2. Magnetic magic

Age range
Six to eight.
Group size
Small groups.
What you need
Thick card, boat with metal attached to bottom, magnet, paper-clip, cotton, Plasticine, stand, magnetic theatre.

What to do
The children can carry out the following three experiments.

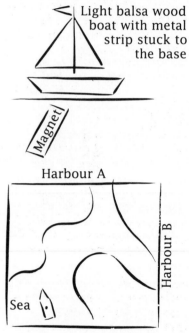

Light balsa wood boat with metal strip stuck to the base

Magnet

Harbour A

Harbour B

Sea

• Using the magnet, the children have to direct the boat into the harbour which is drawn on the card.

• The children can make some cardboard figures and stick them to wooden boxes to which is attached a floating-type ceramic magnet (available from SPO). These figures can then be moved about the theatre by using another magnet so that they appear to be moving by themselves.

• Set up the 'Indian Rope Trick' as in the diagram, and see whether the children can explain what is happening.

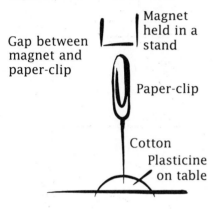

Gap between magnet and paper-clip

Magnet held in a stand

Paper-clip

Cotton

Plasticine on table

Science content
Magnets attract some metals but not others. They will attract iron, steel, nickel and cobalt, but not aluminium, copper, brass or lead.

Further activity
Let the children explore various objects in the classroom (avoiding televisions and video players) to see if they are magnetic. The children could predict and then test to see whether given objects are attracted or not.

Ask the children to explore which materials a magnet will attract through. Does the thickness of the material affect this? Their findings can be related to activities above.

Safety
Do not put magnets near televisions or videos, as this may reduce the picture quality.
AT11/2a

3. Getting your bearings

Age range
Six to eight.
Group size
Small groups.
What you need
Large iron or steel needle, washing-up bowl full of water, drop of washing-up liquid to remove surface tension effects, piece of cork, magnet.
What to do
Can the children work out how they could find their way out of a forest or jungle with only a magnet, a needle and a piece of cork in their pocket?

They should draw the magnet over the needle at least twenty times, in the same direction each time, to magnetise it. The magnet should be taken from the top back down to the bottom of the needle in a wide arc, so as not to demagnetise the needle.

The children should then carefully place the needle on the piece of cork in the bowl of water (to represent a pool in the forest). The needle should point north/south. They can verify this with a compass.
Science content
Children should find out about the Earth's magnetic field using a compass. Magnets attract certain materials but not others, and can repel each other. A compass is of course a magnet.

The atoms in a needle are arranged randomly. When the magnet is passed over the needle the atoms are reorientated so that the needle acts as a magnet and compass. The end of the needle which points north will depend on the pole of the magnet used to magnetise it. In the example given, the pointed end of the needle will face north as the north end of the atoms will be attracted to the south pole of the magnet.

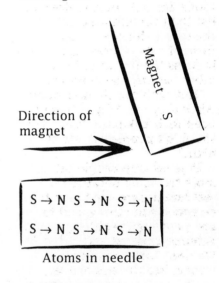

Direction of magnet

Magnet

S

| S → N | S → N | S → N |
| S → N | S → N | S → N |

Atoms in needle

Further activity
Carefully place a magnet near one end of the needle compass without touching it. What happens? Put the other pole of the magnet near the same end of the needle. Like poles (north to north, south to south) will repel, while unlike poles (north to south) attract each other.
Safety
Make sure that care is taken when using needles.
AT11/2a

4. Light it up: getting started with electricity

Age range
Six to nine.
Group size
Pairs.
What you need
One battery, two wires (with or without crocodile clips), one bulb (without bulb holder), photocopiable pages 167 to 169.
What to do
Ask the children to light a bulb using the battery and one wire. Let them investigate until they get the bulb to light. Encourage them to try different ideas.

Give the children some assistance if they appear to get frustrated; for example, tell them to place the bulb on the top or bottom of the battery. When they excitedly get the lamp to light, ask them to tell you exactly where the wires are touching. Talk to them about the passage of electric current around the circuit; an enlarged diagram of a bulb would be useful.

Give the children a further wire and ask them to light the bulb again, using all four pieces of equipment. (They should position the bulb between the ends of the two wires.) Ask the children to show you how the electric current is travelling through the circuit.

Science content
In order for the bulb to light, a circuit must be constructed. This allows the electric current to flow from the battery, through the wire lead and bulb, and back to the battery. It may be helpful for the children to relate 'circuit' to racing circuits in athletics and car racing.

An electric current travels through the base of the lamp (the grey soldered part), up and along the filament, and out through the side of the metal screw thread – or the other way round.

Further activity
Give the children a number of circuits (see photocopiable page 168) and ask them to predict and then test whether the lamp will light.

Discuss the dangers associated with the use of electricity in the home, school and community and the appropriate safety measures which should be taken, such as not switching on appliances with wet hands and avoiding railway lines and pylons.
Safety
You should stress the harmless nature of this equipment, but make sure children are aware of the dangers of mains electricity.
AT11/2b, 3b

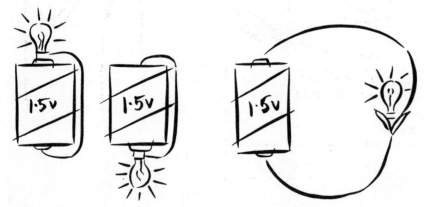

5. Conducting electricity

Age range
Seven to nine.

Group size
Pairs.

What you need
Battery, bulb in a bulb holder, three wire leads with or without crocodile clips, sticky tape, variety of materials including metal discs, wood, plastic, rubber.

What to do
Ask the children to construct a simple circuit with one battery, one bulb and three leads. They should attach the wire from the bulb to the battery with sticky tape or use a battery box.

Science content
Some materials conduct electricity while others do not. Those that conduct electricity (conductors) will allow the circuit to be completed and the bulb will light up. Those that do not conduct electricity (insulators or non-conductors) will not allow the circuit to be completed, and the bulb will not light up.

Further activity
Can the children see any patterns in the results? Many metals conduct electricity, but do all metals? Ask the children to test this. Most non-metals do not conduct electricity, but there are exceptions, for example graphite (carbon).

Safety
The children are using one 1.5 volt battery and harmless. However, it m stressed that the electri-y in the school, in their homes, in the community, on the railway lines and in pylons is very dangerous and can kill. Batteries in cars and electric fences around fields will also give a shock.
AT11/2b, 3a

The children can then place the items, one at a time, so that they touch the two crocodile clips at the ends of the other wires, thus completing the circuit. Does the bulb light up? They should record the results in a table.

Lets electricity pass (bulb lights)	Does not let electricity pass (bulb does not light)
Copper	Rubber

6. Quiz your friends

Age range
Seven to eleven.

Group size
Pairs.

What you need
Piece of thick card, balsa wood or hardboard to use as a quiz board, brass paper fasteners, short lengths of wire (with or without crocodile clips), two pieces of paper for questions and answers, one or two batteries connected to a buzzer or lamp.

What to do
Design a quiz board using electrical circuits. There must be sufficient room for the questions, answers, paper fasteners and wires. An example of a possible design is given below. The paper fasteners could be sited on the margins of the board, but this will use more wire.

First you should decide on the topic for the quiz board and work out how many questions and answers you will have.

Next you should make holes in the board as shown, and push a paper fastener through each hole, bending the fasteners open. Making a good

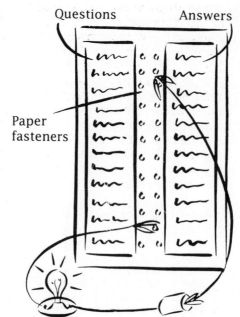

firm connection (either using crocodile clips or by winding the bare wires around the fasteners), you should join up pairs of fasteners with wires as shown. It is best not to join up too many pairs which are opposite each other.

Write your questions on one piece of paper and the answers on another, and arrange them so that questions and answers are linked by the wires behind the board. Then the children can answer the questions using

the battery, lamp or buzzer and wires to complete the circuit. Alternatively the children could design their own quiz boards.

It you use detachable pieces of paper, the quiz board can be used over and over again – until the children remember the connections!

Science content
This activity will help the children to understand that a complete circuit is needed for an electrical device, such as a bulb or buzzer, to work. If they are able to make the game themselves they will have shown that they are able to construct simple circuits. However, they are really applying their knowledge, and there are simpler ways (though perhaps not so fulfilling) to demonstrate whether children are able to construct electrical circuits.

Further activity
The board could be used to link famous scientists and their inventions.
Safety
Ensure that the children are reminded that while the circuits in use in this and other activities are not harmful, those at home and in the community are potentially lethal.
AT11/3b, 4a

7. Drawing your circuit

Age range
Ten to eleven.
Group size
Individuals.
What you need
Photocopiable pages 170 and 171.
What to do
Ask the children to set up a circuit such as the one shown below.

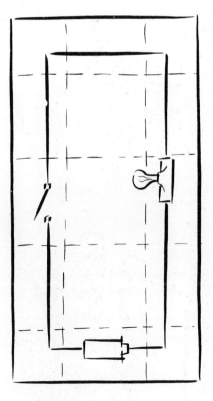

Ask them to draw this circuit in two dimensions. They can also use the cut-out circuit pictures to construct the circuit (see below and photocopiable page 170).

They can then use the circuit diagram symbols on photocopiable page 171 to construct a circuit diagram. For example:

Science content
There are certain internationally accepted symbols for recording electrical circuits diagrammatically. Young children could draw pictures before drawing the diagram above, finally moving on to conventional circuit diagrams. The symbols used are:

Battery 1.5v	▮⊩	Switch (off)	⟍
Batteries (cell) 3v	▮⊩▮⊩	Switch (on)	━●━●━
Bulb	⊗	Wire	━━━
		Buzzer	⌣

Further activity
The children can repeat this activity with other circuits and look at circuit diagrams for electrical appliances.
Safety
These circuits are harmless, but children will need constant reminding that the electrical circuits in their home, school and community are extremely dangerous.
AT11/5a

Doorbell

Doorbell box ringing

Outside   Inside

8. Shadows

Age range
First activity, six to eight; second activity, nine to eleven.

Group size
Small groups.

What you need
Torch (preferably with lens bulb), paper, tissue paper, perspex, acetate sheet, glass or window, door etc.

What to do
Tell the children to put their hands between a torch or overhead projector and a sheet of white paper in a darkened corner. Can they make the shadow smaller or larger? Can they make silhouettes of animals, such as a bird, mouse or rabbit?

Make available some paper, tissue paper, acetate sheets and perspex, and use the glass on the windows. Let the children make shadows on these surfaces. Can they make a shadow in water or in the air without any other substance behind it?

The children can investigate how to change the shape and size of the shadow.

Science content
Light passes through some materials, such as air, and not others, such as paper. When it does not pass through, or not completely, shadows may be formed. Opaque or translucent materials cast shadows. In fact, it is difficult to get substances which do not cast shadows. Even an acetate sheet produces a weak shadow.

The nearer the object is to the light source, the larger the shadow (see diagram below). Light travels in straight lines, and this explains the shape and size of shadows. The shape of the shadow is the same as the outline of the object.

Further activity
Repeat the first activity using head silhouettes. Cut out the silhouettes and pin them around the classroom. Can the children recognise each other's silhouettes?

Let the children make and give a shadow puppet show, with card puppets attached to sticks.

AT15/2a

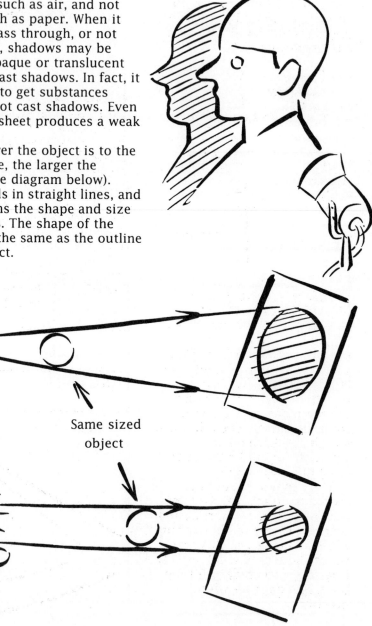

Same sized object

50

9. Changing the direction of light

Age range
Six to eight.
Group size
Small groups.
What you need
Torch, mirror, numbered target.
What to do
Ask the children to use a torch and a mirror to find a way to direct light around a corner of the classroom so that it shines on the corridor wall.

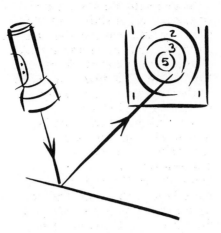

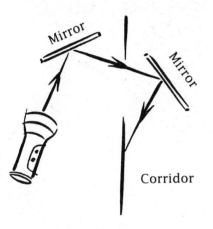

This can be extended by playing the game of 'light shooting'. Attach a numbered target to the wall. Hold the mirror so that it points in the general direction of the target. With the torch switched off, the children should position the torch one or two metres from the mirror. Then they can switch the torch on and see where the light hits the target. What did they score?

The children should then switch the torch off, readjust the position of the mirror, and then switch the torch on again. They should repeat this five times and record their results. Does their accuracy in hitting the target improve with practice?

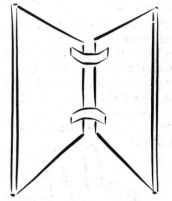

Science content
Light travels in straight lines. Its direction can be changed by using shiny objects like mirrors.

Light falling on a plane mirror is reflected. The angle between the incoming light and the mirror is the same as the angle of the outgoing or reflected light (see Activity 10).

Further activity
Tape two mirrors together. On a piece of paper draw angles of 30°, 60°, 90°, 120° and 150°. The children can then adjust the two mirrors to each of the angles in turn and predict and then test how many reflections of an object, placed between the two mirrors, they will see (see also Activity 10).

Safety
Plastic mirrors should preferably be used instead of glass.
AT15/3a, 3b, 5a

Angle	Predicted number of reflections	Actual number of reflections
30°		
60°		

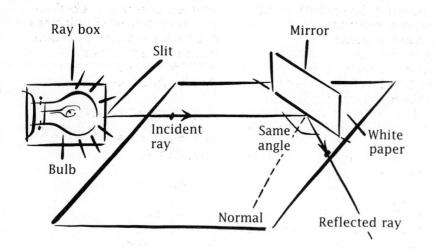

Labels for top diagram: Ray box, Slit, Mirror, Bulb, Incident ray, Same angle, White paper, Normal, Reflected ray

it, or is reflected. The incoming ray (incident ray) always makes the same angle with the mirror as the reflected ray. Thus the angle of incidence always equals the angle of reflection. Strictly speaking, the angles of incidence and reflection are the angles between incoming and outgoing rays and a line at 90° to the mirror, at the point where the ray is reflected. This line is called the *normal*. When the incoming ray is shone along the normal no reflected beam is seen, as it is reflected back along the normal.

Further activity

The children can build a periscope using two plane mirrors so that they can see around corners or over a wall. The light ray is reflected at the same angle as it comes into the mirror so that it is reflected downwards to the second mirror and then into the eye.

Safety

Electrical apparatus should be regularly checked in accordance with the Electricity Work Regulations 1989.

AT15/5a

10. Reflection of light

Age range
Ten to eleven.

Group size
Small groups.

What you need
Ray box or light box, mirror, white paper, pencils.

What to do
Ask the children to follow these instructions.
• Set up the equipment in a darkened corner of the classroom as shown in the diagram above (an Osmiroid light box may be used in place of one of the mains ray boxes).
• Draw a line along the length of the mirror.
• Using a sharp pencil, put a dot where the ray hits the mirror; also draw a dot on the line of the incoming ray and the reflected ray (see diagram).
• Remove the mirror, join up the dots with two straight lines and measure the angle between the mirror and the incoming ray and the angle between the mirror and the reflected ray. Is there any pattern?

• Replace the mirror, but change the angle of the incoming ray by moving the ray/light box.
• Repeat the experiment as before.
• Draw a table of your results.

Science content
A ray box or light box contains a narrow slit to produce a thin beam of light. When this beam meets a mirror it bounces off

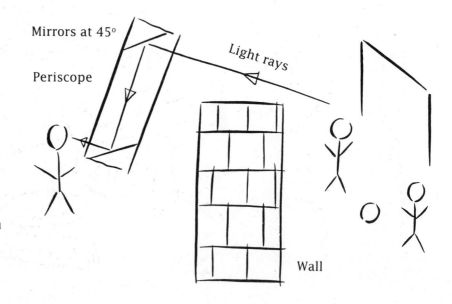

Mirrors at 45°

Periscope

Light rays

Wall

11. Melting and solidifying

Age range
Six to eight; further activity, nine to eleven.
Group size
Pairs or small groups.
What you need
Ice, ice-cream, chocolate, wax, tray, pan, heat source.
What to do
Place some of the ice, ice-cream, chocolate and wax on a tray in the classroom and the rest in a pan on a heat source.

Encourage the children to observe what is happening.
• Can they predict what will happen to the substances, and which will melt first?
• What happens when the substances melt – what do they observe?
• Which melted in the classroom and which needed heat to melt them?

• How did the heat affect the speed of melting?
Science content
Heating or cooling materials can cause them to melt or solidify; it can also change them permanently, as in baking. Different substances melt at different temperatures, but melting and solidifying occur at the same temperature in the same substance. Substances change from solid to liquid on heating (further heating changes them to a gas), and from a liquid to a solid on cooling.
Further activity
Why does chocolate melt in your fingers or your pockets? The children can measure the temperature at which different substances melt, and describe the changes that occur.
Safety
Supervise the use of all heat

sources. Ensure that children do not eat any food which has been used in an investigation on unclean surfaces etc.
AT6/2c, 3b, 4d
AT13/3b, 4d, 4e

12. What is my weight (mass) and volume?

Age range
Nine to eleven.
Group size
Small groups.
What you need
250cm³ measuring cylinder; scales, a number of small items that can be successfully submerged in water; for example, eraser, pencil sharpener, small pencil, pen top.

ICE CREAM   CHOCOLATE   WAX

What to do

Present the small items to the children and ask them what they would do to find the weight (mass) and volume of these regular and irregular shaped objects. The children can then fill the 250cm³ measuring cylinder to the 150cm³ mark. Ask them to weigh one item and record its weight/mass. They should then drop the item into the water making sure that it is completely submerged, and look at the water level. They can then repeat this process with the rest of the objects, recording their results in a table as shown below.

Science content

Solids and liquids have a weight/mass and volume. The volume may be measured and calculated for regular shapes such as cubes, but it is difficult to do so for irregular shapes. The volume of small objects may be found relatively accurately by immersing them in water, as the object displaces its own volume of water, and hence raises the level of the water.

Further activity

Tell the children the story of how Archimedes dropped the soap and the water flowed over the side of his still bath, and how he ran home naked shouting 'Eureka'! The volume of objects may be more accurately calculated using a 'Eureka can', with the displaced water captured and measured in a measuring cylinder.

Safety

Avoid the use of sharp objects.

AT6/4c

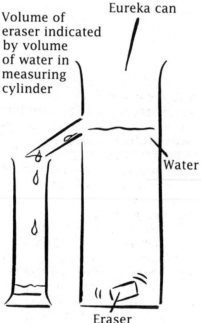

Volume of eraser indicated by volume of water in measuring cylinder

Eureka can

Water

Eraser

Item	Weight/ Mass	Startingvolume of water	Finishing volume of water	Volume of water
Eraser		150cm³	159cm³	9cm³
Pencil sharpener		159cm³		

13. Solids, liquids and gases

Age range
Nine to eleven.
Group size
Small groups.
What you need
Block of wood, metal cube or strip, small measuring cylinder, water, three syringes (one containing a small piece of dowel, one containing a small volume of water, and one containing air), and boiling kettle, or pan of boiling water, or water in a test tube over a heat source.
What to do
Ask the children to list some items in the classroom and outside and put them in three sets: solids, liquids and gases.

Discuss and consider with the children the properties of solids, liquids and gases and draw up a table to illustrate their similarities and differences. Questions such as the following may be posed:
• Are solids, liquids and gases generally hard or soft?
• Can the shape and volume of solids, liquids and gases be easily changed?

Place a biro top in a measuring cylinder and then 'pour' it into a beaker and then on to the bench. Has its shape changed? Has its volume changed? Repeat using a small volume of water. Has its shape and volume changed?
• Taking safety precautions, let the children observe some boiling water and the steam being produced. Is the shape of the gas changing?

Using the three syringes, the children can try to squeeze the solid (wood), liquid (water) and gas (air). Which can be compressed? Which can be changed in volume?

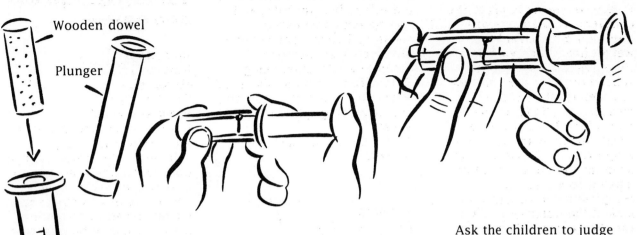

Wooden dowel

Plunger

The children should put the wood in the syringe and try to squeeze it with the plunger. They can then put 5cm³ of water in the syringe. With a finger tightly over the end, they can try to squeeze the water. Next, they should lower the plunger so that there is 5cm³ of air in the syringe and, with a finger tightly over the end, try to squeeze the air.

Science content
Materials may be classified into solids, liquids and gases on the basis of their properties as illustrated below, although some materials and substances do not fit easily into these categories.

Further activity
By heating ice to become water and then steam, and then cooling steam to become water and freezing water to become ice, children will be helped to understand the sequence of changes of state that result from heating and cooling.

Ask the children to judge whether jelly, powder, Plasticine and sponge are solids, liquids or gases. Ask them to explain their answers.

Safety
Care must be taken with all types of heating appliances.
AT6/4d, 4e

14. Is it hot or cold?

Age range
Five to seven.

Group size
Small groups.

What you need
Thermometer (–10°C to 110°C), three bowls of water (one hot, one lukewarm and one ice-cold), fridge.

What to do
Go around the school with the children, touching various items (with safety in mind) and recording whether they are hot, cold or lukewarm. Take the temperatures of some of these things, and record them as in the table on page 56.

Back in class, the children can take turns to put one hand in the bowl of ice-cold water and one in the bowl of hot water, and then put both hands in the lukewarm water. How do their hands feel? The one that was in the hot water should feel cold, and vice versa.

State of matter	Hard/ soft	Can its shape be changed?	Can its volume be changed?	Can it be compressed?
Solid	Hard	No	No	No
Liquid	Soft	Yes	No	No
Gas	Soft	Yes	Yes	Yes

Item	Did it feel hot, lukewarm or cold?	Temperature
Fridge	Cold	4°C (low)
Radiator	Hot	
Wall	Lukewarm	
Air outside	Cold	
Air inside	Lukewarm	

Science content

Temperature is a measure of how hot or cold things are. Heat is the amount of energy in a substance; for example, a large cup of water at 30°C has the same temperature but more heat energy than a small cup of water at 30°C, as a larger volume needs more energy to keep its temperature than a smaller volume. Some things feel cold to our bodies, some warm and others hot. Sometimes our senses are confused, as in the lukewarm water above. That is why we need thermometers to measure how hot or cold things are.

Further activity

Repeat the air temperature readings at intervals; for example, once a term, once a month or once a day for a set period. The children could take temperature readings with a thermometer as well as by feel.

Safety

Do not use water which is too hot to the touch! Make sure that the thermometer is handled carefully; spirit-based thermometers are preferable to mercury ones. Wrap an elastic band around the thermometer to stop it rolling, if it does not have a triangular collar.

AT13/4d

Cold water   Lukewarm water   Warm water

15. Belt up, transfer the energy

Age range
Seven to eleven.
Group size
Pairs.
What you need
Materials to build a vehicle using Jinks Construction Techniques, for example 1cm pine or Malayan jelutong wood, dowel, cardboard, junior hacksaw, photocopiable pages 172 and 173, glue gun or strong adhesive, 1.5V battery, battery box, two wires, small pulley, elastic band, plastic tubing washers or grommets between wheels and chassis.
What to do
Ask the children to construct a buggy using the instructions on photocopiable pages 172 and 173. This could involve precise measurements for the wood used (minimum 20cm × 9cm). Ensure that the elastic band is tight against the drive pulley and the small pulley attached to the motor, so that the axle rotates. Then stick it to the chassis using the hardboard. Too much tension on the elastic band will increase the friction so that the axle will not rotate; too little tension on the elastic band will decrease the friction to such an extent that the axle will not rotate. It is important that the motor is rigidly attached to the buggy, or vibration will prevent movement of the axle.
Science content
Chemical energy in the battery produces electrical energy. This electrical energy is converted into movement energy in the motor which turns the spindle. This movement energy is transferred from the motor to the axle by the belt (elastic

band) which in turn makes the wheels rotate.

Further activity
• Can the speed of the vehicle be increased? For example, use two batteries.
• Can the vehicle be made to move quicker? For example, use wooden or plastic wheels.
• Use wheels made of a single thickness of card. How does this affect the movement?

Safety
Glue guns convert solid adhesive into molten adhesive and it sets quickly and firmly on cooling. However, the glue gun itself may get hot and molten adhesive could be 'flicked' across the classroom. Use of glue guns is prohibited by some local authorities.
AT11/5b

16. Into gear

Age range
Seven to nine.
Group size
Small groups.
What you will need
Home made gears: bottle tops; corrugated card around lids; cocktail sticks or dowel in a circle of potato or polystyrene – or bought gears such as 'First Gear', which is available from SPO and Heron Educational Limited.

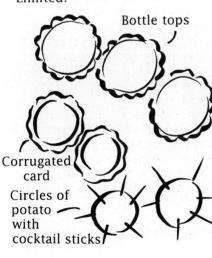

Bottle tops

Corrugated card

Circles of potato with cocktail sticks

No. of turns small cog	No. of teeth	No. of turns large cog	No. of teeth
10			

What to do
Pin the cog wheels to a board so that the teeth interlock. Ask the children to turn one of the cogs and note which way the other cog(s) move. They should then mark the cogs in some way and turn the smallest cog to complete ten turns. Other members of the group could count the number of turns of each cog. The children can record the results and look for a pattern in the numbers, relating the number of turns to the number of teeth.

Science content
One way in which energy may be transferred from one place to another is by using cogs or gears. Cogs or gears can change the direction of movement, and can speed up or slow down the movement, depending on the number of cogs, or change the plane of movement.

Further activity
Ask the children to look at the way cogs or gears may be used in a windmill or watermill to change the direction of movement, so that the mill can grind the corn.

Safety
Bottle tops may be sharp and care should be taken when pinning them to a board.
AT13/3c

Vertical

Horizontal

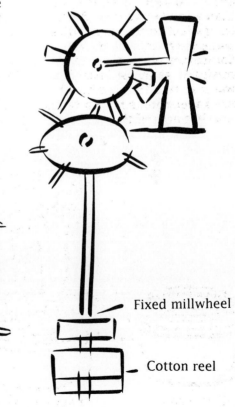

Fixed millwheel

Cotton reel

17. The balloon monorail

Age range
Nine to eleven.
Group size
Small groups.
What you need
Balloons, pump, cotton, plastic drinking straw, sticky tape.
What to do
Ask the children to pump up a balloon, and then release it. Does it go far? Does it go in the same direction each time? Does a round balloon travel in the same way as a long one?

The children can then set up the apparatus shown in the diagram below. They should thread the drinking straw on the cotton, and fix the cotton across a corner of the classroom. Then they can attach the deflated balloon to the straw with sticky tape, and blow into the balloon, pinching the neck between their fingers until they are ready to let go. How many puffs are needed to make the balloon travel one metre, five metres etc?
Science content
Energy is stored in the balloon. Movement is in the opposite direction to the thrust of air escaping. A similar phenomenon causes movement in a space rocket or jet engine.
Further activities
See Activities 18 and 19.

Safety
A cotton or gut line is almost invisible and is potentially dangerous. Set it up in the corner of the classroom, above head height.
AT13/3a, 3c, 4c

18. The spring dragster

Age range
Nine to eleven.
Group size
Pairs.
What you need
Card, photocopiable pages 174 and 175, spring, two pieces of 4mm diameter dowel (18cm and 12cm long), one piece of 6mm diameter dowel (18cm long), three plastic cotton reels, paper fastener, sheet of A5 paper, four rubber grommets or pieces of plastic tubing or rolled-up elastic bands, cork or rubber bung, adhesive tape.
What to do
The children should cut out the body of the spring dragster from the photocopiable pages, and stick it to a piece of card 1mm thick. They can then follow the instructions to build the vehicle. Once they have built it they should:
• make it move a short way;
• make it go further;
• modify it to make it travel further still;

• graduate the large piece of dowel attached to the spring to see how far the buggy travels with one unit of force, four units of force etc;
• draw a table or graph to express their results.
Science content
Energy may be stored in the spring and later released to produce movement energy via the wheels. The movement of the dragster depends on the size and direction of the forces exerted on it.

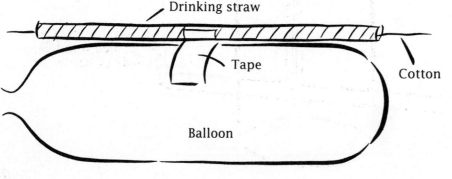

Drinking straw

Tape

Cotton

Balloon

NB Children should be encouraged to repeat their trials three to five times and take an average to improve the validity of their results.

Further activity
Ask the children to investigate on which surfaces the dragster travels furthest for the same extension of the spring (unit of force). This introduces the idea of friction, the force one object meets when it moves over another.

Safety
Take care to observe normal safety procedures when children are involved in construction.

AT10/3a, 4a, 5d
AT13/3a, 3c, 4c

19. The bottle roller

Age range
Nine to eleven.

Group size
Pairs.

What you need
Washing-up liquid bottle, two pieces of dowel or cane, smooth card or plastic square, large elastic band, wire coat hanger.

What to do
Let the children construct the roller as shown in the diagram. To do this, they should cut the end off the washing-up liquid bottle and, holding an elastic band at the top of the bottle with a piece of dowel, thread it through the top of the bottle and then hold it in place with another piece of dowel which has been pushed through the sides of the bottle. They may find that a wire coat hanger will be useful to pull the elastic band from the nozzle to the lower piece of dowel.

Once they have constructed the roller they should consider and predict how it will move. Then they can experiment with ways of making the roller go further or faster by changing one variable, for example the number of turns or the size of the elastic band.

Science content
The key idea in this activity is that energy may be stored and used later. When the elastic band is wound up, energy is transferred from the body into the elastic band. This wound elastic band now contains energy, and as it has the potential to do something, it is called potential energy. When the cane is released, the energy in the elastic band is concentrated into movement energy, or kinetic energy, and some is converted into heat because of friction.

Further activity
Make a spring dragster as shown on page 58, or organise a washing-up liquid bottle roller race!

Safety
Observe normal safety procedures when children are involved in construction.

AT13/3a, 3c, 4c

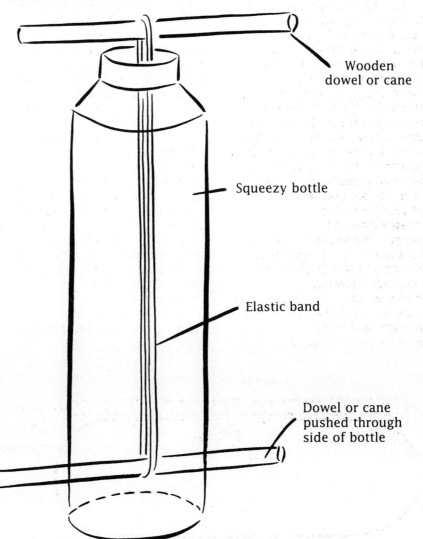

Wooden dowel or cane

Squeezy bottle

Elastic band

Dowel or cane pushed through side of bottle

CHAPTER 5

Sounds around us

Sound is very important in everyday life. The children's day is full of a wide variety of sounds, but how often do they stop to analyse the different types of sound and how these sounds are made? Why do we like some sounds, whereas others can be unpleasant and even frightening? This topic provides the opportunity for children to create their own sounds and delve into the world of music and musical instruments. Many opportunities should be provided for listening to sounds from the environment, thus helping children become aware of the enormous range of sounds which are part of their lives.

Children should investigate ways of making and experiencing sounds. They should explore the changes in pitch, loudness and timbre of a sound and be made aware that sound can be pleasant or obtrusive in the environment.

ACTIVITIES

1. How many sounds can you recognise?

Age range
Five to seven.
Group size
Pairs.
What you need
Blindfold, range of junk materials such as elastic bands, shoe box, small blocks of wood, tin containing beads, plastic bottle, wooden drumstick etc.
What to do
How many sounds can the children recognise? Ask one of the children in the pair to put on a blindfold while the other makes some different sounds with the junk materials. The child wearing the blindfold should try and guess how the sounds are being made.

The children should try to make a lot of different sounds in different ways, such as by shaking the tin, scraping the bottle with the drumstick, and twanging the elastic bands.

Science content
Sounds can be made in a variety of ways, such as plucking, shaking, scraping or blowing.
Further activity
Investigate musical instruments, for example drums, guitars, recorders and xylophones. Children can try to classify the way sounds are produced.

Make an audio tape of sounds for a 'sounds quiz'.
AT14/1a

2. Sounding out

Age range
Seven to nine.
Group size
Small groups.
What you need
Elastic bands of different thicknesses around different sized boxes, straws, plastic bottle, wooden stick, tins, wooden beads, water.
What to do
Let the children use the materials provided to make sounds in different ways. They can blow down the straws,

scrape the side of the bottle with the stick, tap the side of the tin, make a shaker using a tin containing wooden beads, and twang the elastic bands.
Science content
Sounds can be made in a number of different ways; for example by plucking, blowing, scraping, banging, and by using different materials.
Further activity
Use each method to produce a number of different notes; for example, try using different sized elastic bands on one box. Can the children now make low sounds? Can they make high sounds? Can they put the sounds in order? See if the children can make some musical instruments out of junk materials, such as a reed pipe, made by cutting the end of a straw into a V-shaped wedge which is placed in the mouth and blown. A shorter straw gives a higher note.
AT14/1a, 2b

3. Make a bottle organ

Age range
Seven to nine.
Group size
Small groups.
What you need
Eight empty milk bottles (use identical bottles).
What to do
Sounds can be made by blowing over the top of a milk bottle. Place the bottles in a row. Put different amounts of water in the bottles. Ask the children to blow across the top

of each bottle to make a sound. Can they make a musical scale? Can they play a tune on the bottle organ?

Science content
Blowing across the top of a bottle makes the air vibrate inside the bottle. Different amounts of water in the bottles produce different musical notes. Organs depend on air vibrating in pipes of different lengths.

Further activity
Instead of blowing across the top of the bottles, the children could *gently* tap the bottles below and above the water level.

Safety
Care needs to be taken when handling glass bottles.
AT14/2b

4. What is sound?

Age range
Seven to eleven.
Group size
Individuals.
What you need
Ruler.
What to do
Ask the children to hold a ruler across the edge of a table, press one end firmly to the table and flick the other end. What can they see? What can they hear? Next they can move the ruler so that there is less of it hanging over the edge of the table. Do they notice any differences when they flick the ruler?

Science content
Anything which vibrates produces a sound. The part of the ruler hanging over the edge of the table vibrates when it is tapped, and a sound is produced. Shortening the vibrating section causes an increase in speed of the vibrations and thus a higher sound is heard (that is, the

pitch changes). The pitch of a sound depends upon the number of vibrations per second. Tapping harder makes the vibration stronger and produces a louder sound.

High-pitched notes have a high frequency and low-pitched notes have a low frequency. Frequency is measured by the number of vibrations per second, with a unit called a Hertz (Hz). The lowest notes the human ear can pick up have a frequency of about 15 to 20 Hertz, and the highest notes that we can detect are around 15,000 to 20,000 Hertz.
AT14/3a, 5a, 5b

5. Telephone talk

Age range
Seven to nine.
Group size
Pairs.
What you need
Two yoghurt pots, two matchsticks, a long length of thin string or wire.

What to do
The children should make a hole in the bottom of each yoghurt pot, thread the ends of the string or wire through each hole and tie them to a matchstick. The children can then take one of the pots each, move apart until the string is pulled tight, and try speaking into one yoghurt pot and listening through the other one.

Science content
Sound waves produced by the voice travel down the string or wire as long as the string is held taut.

Further activity
Will the telephone work around corners?
AT14/3a

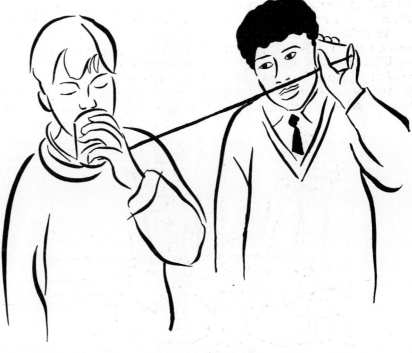

6. Good vibrations

Age range
Seven to nine.
Group size
Small groups.
What you need
Tuning fork, bowl of water, assortment of soft and hard materials, cushion, polystyrene, towel, blocks of wood, foam rubber, cork.
What to do
Ask the children to strike the prongs of a tuning fork on a hard object such as a desk, and watch the prongs vibrate. Then they can put the ends of the fork on the surface of the water in the bowl. What do they see? What do they hear?

They should then make the prongs of the fork vibrate again by striking it on a hard object, and place the base of the fork on different materials in turn to make a sound. When do they hear the loudest sound?
Science content
When a tuning fork is struck and placed on a surface, it causes the surface to vibrate. Softer materials absorb the vibrations and so the sound is quieter. Harder materials like metal produce a louder sound.
Further activity
The children could investigate how sound travels through wood or metal by pressing their ears to the desk and tapping it. Working in pairs, they could investigate how sound travels along iron railings in the playground.
AT14/3b

7. Making waves

Age range
Nine to eleven.
Group size
Pairs.
What you need
A slinky spring.
What to do
Stretch a long slinky spring along a smooth surface. The children can see how part of the spring is stretched and part is squashed when they shake one end of the spring.

Science content
Sound travels in waves. These waves are caused by the air vibrating. A sound wave travels through air like the wave passing down the slinky spring. Some parts of the spring are squashed and pushed closer together (compression) and some parts are stretched and pulled further apart (rarefaction). So with sound waves, vibrations make the air through which they travel stretch and squash.
AT14/3b

8. Make a quiet hat

Age range
Seven to eleven.
Group size
Pairs.
What you need
A range of junk materials such as paper, cotton wool, different fabrics, newspaper, polystyrene, adhesive, sticky tape.
What to do
Ask the children to use the materials they are supplied with to design and make a 'quiet hat' which will keep out noise.

They should then test their hats to see if they really keep out the noise. Let one member of each pair wear his partner's hat while the partner drops a pen to the floor a couple of metres away, or makes a similar small noise. Can the wearer of the hat hear the noise? Which materials do the children think make the 'quietest' hat?
Science content
Soft materials such as fabrics absorb the vibrations which cause sounds.
Further activity
Relate this work to noise pollution and sound insulation.
AT14/3b, 5c

9. Does sound take time to travel?

Age range
Nine to eleven.
Group size
Small groups.
What you need
Digital stop-watch, metre ruler or tape measure.
What to do
Ask the children to measure out a distance of at least 30 metres from a high wall, and then stand there. They should then clap their hands together once and listen for the echo. With a digital stop-watch they should time how long it takes to hear the echo.

$$\text{Speed of sound} = \frac{\text{Distance to wall} \times 2}{\text{Time recorded on stop-watch}}$$

Science content
The sound travels to the wall and is reflected back again to the ears. The sound is heard again a few seconds later as an echo.

$$\text{Speed} = \frac{\text{Total distance sound travels (m)}}{\text{Time (secs)}}$$

The speed of sound in air is approximately 330 metres per second.

Further activity
• Relate this activity to thunder and lightning. You see the lightning first and hear the thunder a few seconds later, because the speed of light is greater than the speed of sound.
• Talk about bats using echo location.
AT14/3b

10. Sounds around!

Age range
Nine to eleven.
Group size
Small groups.
What you need
Noise meter.
What to do
Ask the children to make a noise map of the school by using the noise meter to record the noise level in different places and at different times.

Alternatively they could make their own scale of noise levels, like that shown below.

0	No noise
1	Pin dropping
2	Pencil dropping
3	Clapping hands
4	Door slamming
5	Aeroplane passing overhead

The children can then use the scale to make their noise map of the school.

Science content
The loudness of sound is measured in decibels (dB). The faintest sound discernible to the human ear is about 1dB. A sound loud enough to cause pain would be at a level of about 120dB.

Further activity
Discuss with the children improvements that could be made to control the noise level in the school, such as putting carpets on the floor, and keeping doors closed. Initiate discussion on noise control in the environment, for example in the vicinity of airports, and talk about the use of double glazing, ear muffs etc.
AT14/5c

CHAPTER 6

Weather and seasons

The weather is of great importance in our everyday lives. Talking about it is a famous national pastime which is joked about throughout the world. We have had plenty to talk about in recent years, however, as mild winter has followed mild winter, long hot summers have followed long hot summers and between them we have enjoyed early springs and late autumns.

Through studying the weather, the children will develop their knowledge and understanding of the atmosphere, rocks and soil and the changes these things undergo. From starting school they should be encouraged to observe and record the changes in the weather and to relate these to their everyday activities and to seasonal changes. Young children should start to understand that the weather has a powerful effect on people's lives, and while they might consider it serious that their sledge was little used in the winter, this is nothing compared to the damage caused by winds or floods, or the havoc caused by drought and starvation. Older children should have the chance to make regular, quantitative records of the weather and to make their own weather instruments.

The weather has a daily influence on the landscape and buildings, which should be studied at first hand. Buildings and gravestones are damaged through weathering, and in a similar way soils are formed from the rocks below them.

ACTIVITIES

1. Weather charts

Age range
Five to seven.

Group size
Small groups.

What you need
A card or paper weather chart, cards with the days of the week, months, dates and weather type written on them, cards with pictorial representations of the weather, Blu-Tack.

What to do
Encourage the children to record the weather at the same time each day. Records may be made on a daily, weekly or monthly basis. Involve the children in the construction of the charts as much as possible.

Older children could be encouraged to make charts with more than one symbol, for example:

The temperature and wind directions could also be added.

The daily chart requires cards with the day, date, weather and related picture to be attached with Blu-Tack.

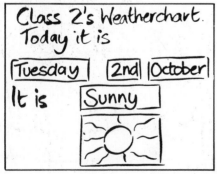

Monday	Tuesday	Wednesday	Thursday	Friday

The children could fill in a weekly chart with symbols they have produced co-operatively. Words could be added to this if they wish, while a monthly chart could record the number of days a particular weather type was experienced. Pupils could record one or more types of weather each day, such as wet and windy, cold and rainy or warm and sunny.

Science content
There are many different weather conditions, which are generally undergoing constant change. The weather may be recorded over a period of time using a variety of methods of communication, for example, using words, drawings, pictures and charts.

Further activity
Collect weather sayings, such as 'Red sky at night, shepherd's delight' and 'The north wind shall blow and it will bring snow'.

AT9/1a, 1b, 2a

2. Weather effects

Age range
Five to seven.

Group size
Whole class.

What you need
Pictures of sunny, rainy, snowy, windy and frosty days.

What to do
Raise the following questions with the class:
• What do you like to drink on a hot day? On a cold day?
• What do you like to eat on a hot day? On a cold day?
• When does the weather make you feel tired?
• When does the weather keep you in?
• When does the weather encourage you to go outside?
• Who might have to work longer when it is icy and snowy?
• Who might not be able to work when it is icy and snowy?
• Who might be pleased because it is snowy/windy/hot?
• How might a farmer be affected by the weather?
• How might a sportsperson be affected by the weather?
• How might sun, rain, wind, fog or snow affect someone's life?

Science content
The weather has a powerful effect on people's lives. It costs the country a great deal of money in the winter to grit the roads, and to pay for vehicle accident damage and burst pipes. Drought may cause crops to fail and so prices rise. Floods can ruin people's houses. Gales blow down trees, fences and sheds and damage buildings.

Further activity
Ask the children what the spring, summer, autumn and winter mean to them. What do they associate with the seasons? When is it hot? When is it cold, frosty, snowy? Try to help them to appreciate that there are patterns in the weather which are related to the seasonal changes.
AT9/2a, 2b

3. Weathering heights!

Age range
Seven to nine.

Group size
Small groups.

What you need
Photocopiable pages 176 and 177, a camera (optional).

What to do
On a rural walk, look for evidence of weathering (see photocopiable pages 176 and 177), bearing in mind that this evidence is extremely difficult to see except for pathway erosion, rock falls and scree slopes.

On an urban walk look at statues, gravestones, steps and old buildings for evidence of erosion. Are there any signs of weathering? Is it worse on one side? Why might that be? Is one type of stone affected more than another?

The children can record evidence on weathering in the form of pictures and writing, and compare these with the pictures on photocopiable pages 176 and 177.

Science content
The slow breaking down of buildings and the landscape is called weathering. Weathering takes place in a number of ways, but wind and rain are the most obvious causes. As well as the physical impact of wind and rain which cause weathering, the wind might carry small particles which further the process on impact. Pollutants in the atmosphere, such as carbon dioxide and sulphur dioxide, combine with water to form weak acids which add to the weathering of stone buildings, rocks and cliffs.

Heat from the sun or from fire and cold in the form of frost cause expansion and contraction of rocks, which eventually causes them to crack, thus allowing more surface area to be eroded by wind and rain.

The roots of plants can also cause rocks to crack as they grow.

AT9/3a, 3c

4. Rocks and soil: is there a connection?

Age range
Seven to nine.
Group size
Small groups.
What you need
Various rocks such as limestone, chalk, sandstone, granite, slate etc, sandpaper, file, containers, safety glasses, magnifying or hand lenses.
What to do
After a discussion about the rocks and soil in their gardens or around the school, the children can collect a rock and rub it with another rock of the same type, or with sandpaper or a file. Using magnifying or hand lenses, what do the children observe?

Ask the children to record their findings in a poem, diagram, picture, prose or a play. Relate the particles so produced by rubbing the rock to the process of soil formation – the children made a sort of soil in five minutes, but nature takes much longer.

Science content
The weathering of rock leads to the formation of different types of soil. Some soils are bright orange-red or yellow due to the underlying sandstone. In the south of England, the whiter soils reflect the chalk deposits below them. In other cases, the link between soil and rock is not so striking.
Further activity
• Together with the children, observe the composition of soil by mixing it with water in a clear container. Then shake it up and allow it to settle.

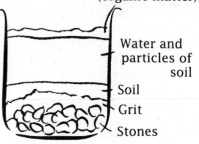

Humus (organic matter)

Water and particles of soil

Soil

Grit

Stones

• Compare the drainage speed of two different types of soils, for example, sandy soil and clay soil.

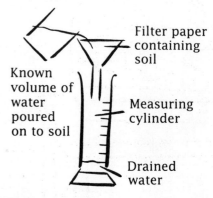

Filter paper containing soil

Known volume of water poured on to soil

Measuring cylinder

Drained water

Safety
Safety glasses are recommended to reduce the risk of small particles of dust getting into children's eyes.

AT9/3c, 3d

5. The weather forecast

Age range
Eight to ten.
Group size
Pairs.
What you need
Map of the British Isles, weather symbols (see photocopiable pages 178 and 179), a video of a weather forecast (useful but not essential).
What to do
Ask the children to record the weather in detail for one day.

Using an enlarged map of the British Isles and the enlarged cut-outs from photocopiable pages 178 and 179, they can then select the appropriate symbols to present a weather forecast to the group or the whole class. For example:

'After a cold, frosty start it will brighten up over Albert Primary School, becoming cloudy in the afternoon...'.

Science content
The information on which the weather forecast is based has been obtained from weather ships, aircraft, local weather stations and, more recently, satellites.
Further activity
Encourage the children to watch the weather forecast on the television or on video. Collect some weather forecasts from the newspapers and display them. Are the forecasts accurate?
AT9/3e

6. Comparing weather forecasts

Age range
Eight to ten.
Group size
Small groups.
What you need
Atlas, photocopiable pages 180 and 181.
What to do
Test the children's knowledge of the meteorological symbols used in the media with a series of questions relating to the two weather charts (see photocopiable pages 180 and 181). You could modify the charts to include symbols which represent the weather conditions near your school.

Questions might include:
• What is the highest temperature on day two?
• What is the lowest temperature on day two?
• Was there any rain on day two?
• Was there any rain on day one? Where? (Use an atlas).
• Was there any sun on day one? Where?
• What was the maximum wind speed on day one and day two?
• Was there any fog or mist on either day? If so, where was it located?
• Which seasons of the year are likely to be represented by day one and day two?

Science content
Children will be helped to understand and interpret common meteorological symbols as used in the media and to appreciate that there are patterns in the weather related to seasonal changes.
Further activity
Encourage the children to watch the weather forecast on the television and to look at the forecasts in the newspapers. Are they accurate?
AT9/2a, 3e

7. Making a wind vane

Age range
Nine to eleven.
Group size
Small groups.
What you need
Dowel, thin piece of balsa wood (30cm × 1.5cm) with a hole larger than the diameter of the dowel, polystyrene tile, two beads, felt-tipped pens, container of soil, compass.
What to do
Ask the children to make the wind vane shown below. It is important that the force of friction between the balsa wood, dowel and beads is reduced to a minimum. They should mark north, east, south and west on the container.

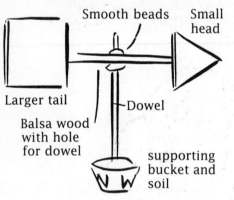

Smooth beads
Small head
Larger tail
Dowel
Balsa wood with hole for dowel
supporting bucket and soil

The children should set up the vane away from buildings, ensuring that the north mark on the vane faces the magnetic north on the compass.

Ask them to record the wind direction at the same time each day, using an appropriate method such as the table shown below.

Day	Date	Wind direction

Science content
Winds get their names from the direction they blow from. Thus a north wind is generally cold, while a south wind is generally warm. Our most common winds are from the west, and they travel across the Atlantic before reaching our shores.

The arrow on a wind vane points in the direction from which the wind is coming. The force of friction must be reduced if the wind vane is to move appropriately.
Further activity
The children could make a wind sock, or design and make their own wind vane.

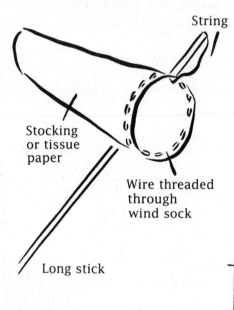

String
Stocking or tissue paper
Wire threaded through wind sock
Long stick

Safety
Any form of construction has possible dangers, though the materials used to construct the wind vane should not cause concern if used sensibly.
AT9/4a

8. Measuring rainfall

Age range
Nine to eleven.
Group size
Small groups.
What you need
Two plastic bottles with straight sides and flat bottoms, scissors or junior hacksaw, funnel and beaker of same diameter, commercially produced rain gauge.
What to do
The children could make a rain gauge in either of the following ways.
• Cut off the top of a plastic bottle and invert the top to give a funnel. Mark a scale on the side of the bottle. Place the rain gauge in a pre-prepared hole (sink it so that the top is 10cm above the ground) or in a container, filling the space around it with sand or soil for stability. Each day, use a ruler or the scale marked on the bottle to record the depth of water collected in millimetres.

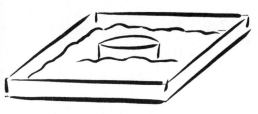

10cm

• Cut off the top of the plastic bottle and place a polythene filter funnel in the remaining part. Secure if necessary. Position the rain gauge as before. Pour the water collected (without spillage) into a beaker with the same diameter as the funnel. As early as possible each morning, record the depth of water collected, using a ruler or a scale marked on the beaker.

Science content

A cloud contains millions of tiny water droplets. When the cloud rises and is cooled, the tiny drops coalesce to form larger droplets which fall as rain. Snow falls if the droplets freeze.

Very accurate measurements of rainfall may be made with purchased rain gauges, as these have been calibrated. With the home-made versions, accurate measurements may only be made if the surface area for collection is the same as the surface area for storage and measurement of the rain, or if calibration is carried out, which would be difficult for most primary school children to understand. Unfortunately, plastic bottles usually have a rim or bulge in the bottom which reduces the accuracy unless *comparative* records only are required. The second type of home-made gauge allows for accurate results as the funnel and beaker have the same diameters and the beaker has straight sides.

Further activity

• Keep records from previous years and let the children compare them.
• Comparative or direct readings for the week could be displayed in corked test tubes or other similar containers, with food colouring added.
• More able children could calibrate their own rain gauges.

AT9/4a

9. Measuring temperature

Age range
Nine to eleven.

Group size
Small groups.

What you need
Spirit thermometer (–10° to 110°C), photocopiable page 182.

What to do
Ask the children to take the temperature:
• inside a fridge;
• in the classroom in shade;
• in the classroom in the sun;
• above a radiator;
• outside in shade;
• outside in the open.

The children should use these readings to find out the warmest and the coldest place. They can use photocopiable page 182 to practise taking more temperatures and relating these to hot, warm and cold.

The children should record the temperature outside in the shade for a month and put the results in a table, bar chart or graph, as appropriate. Encourage the use of a fair test; for example, use the same thermometer at the same time each day, for the same length of time before the reading is taken.

The temperature should be taken with the thermometer still in the area under investigation. Many children withdraw the thermometer before taking a reading, which inevitably leads to inaccuracies.

Science content
Anders Celsius, a Swedish astronomer, suggested the centigrade scale in 1742. The Celsius scale (previously termed centigrade) records the temperature of melting ice at 0° and the temperature of boiling water at 100°. This replaces the scale proposed by a German, Gabriel Fahrenheit, in 1714. He recorded the temperature of melting ice as 32°F and the temperature of boiling water as 212°F.

Further activity
• Introduce the children to other thermometers, for example, a wall or office thermometer, a clinical thermometer, or a thermostik (Osmiroid). Use a thermometer to take the temperature of the soil surface, 5cm down and then 10cm down etc. Do two thermometers record the same temperature?
• Hat-makers used to use mercury, which is used in some thermometers, in making their hats. Mercury is poisonous, and caused many hatters to go mad; hence the expression 'as mad as a hatter'.

Safety
Purchase a spirit thermometer (–10°C to 110°C) with a strong bulb and a triangular cross-section with a triangular collar to prevent rolling. Mercury thermometers should be avoided, although problems only arise when thermometers break and the mercury vaporises. Mercury spillages should be cleared up immediately. Do not let children play with mercury. Thermometers are easily broken, and the glass is a potential hazard. Safe, sensible usage must be impressed upon all children.

AT9/4a
AT13/2a

10. How fast is the wind blowing?

Age Range
Five to seven (first activity) and nine to eleven (second activity).

Group size
Small groups.

What you need
Coat-hanger, tissue paper, paper, thin card, thick card, sticky tape, wood or thin card, strong wire or thin dowel, thin wire or string, photocopiable page 183.

What to do
• Ask the children to construct an anemometer as shown below, sticking the sheets of tissue paper (A), paper (B), thin card (C) and thick card (D) around the wire. For a fair test, ensure that the pieces of paper are the same size. Hold or hang it in the playground.

Clothes hanger

• Older children could construct an anemometer as shown opposite, including an orbiting wind scale. This could be calibrated using the Beaufort scale.

Ask the children to record the wind speed by measurement using the second anemometer, or by seeing which strips move and by how much in the first.

Science content
An anemometer is a device to measure wind or air speed. The Beaufort scale was invented in 1805 by Admiral Sir Francis Beaufort, to estimate wind speed. The original scale was for use at sea, but it has been adapted for use on land. He devised a scale of 13 points (0-12), and at each point he gave the wind a number, a description, a range of speeds and a description of its effects. The Beaufort wind scale is shown on photocopiable page 183.

Further activity
• Can the children design another type of anemometer? For example, they could hang up a table tennis ball and see what happens to it in the wind.
• The children could produce an illustrated Beaufort scale, or find out facts about the wind speeds in Britain and the world in the *Guinness Book of Weather Facts and Feats* (Guinness Publications). See how much they can find out about wind power.
AT9/4a

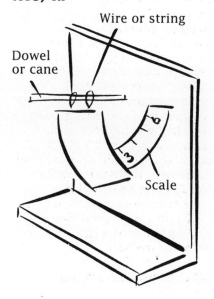

Wire or string

Dowel or cane

Scale

11. Farming and the weather

Age range
Nine to eleven.

Group size
Small groups or the whole class.

What you need
Pictures from newspapers of drought, floods, snow storms, blizzards and drifting.

What to do
On a wet day, ask the children how they feel about the weather. How might gardeners or farmers feel about it? If it is harvest time for cereals they might agree with the children, but if there is a drought or the crops are growing they would probably disagree.

How might wet or windy weather affect farmers at harvest time? How might prolonged freezing weather affect a farmer in January to March, which is lambing time? How would a late winter or early spring frost affect fruit farmers? How might drought affect the size and yield of a crop such as cereals, potatoes or apples?

Broaden the issue to include recent catastrophes brought about by weather conditions, such as flooding in Bangladesh or droughts in Ethiopia. What are the implications when the whole crop is destroyed and the country is already poor, by our standards? The children could relate this work to pictures from newspapers or other sources.

Science content
Children should learn that climate determines the success of agriculture. They should understand the impact of occasional catastrophic events on the communities affected.
AT9/2b, 4b

CHAPTER 7

Human influences on the earth

Children should develop knowledge and understanding of the ways in which human activities affect the earth. Many children will already be very aware of the need to look after our environment. They will have seen bottle banks and collecting points for aluminium cans, and environmentally-friendly products for sale in supermarkets. Many children will have seen their parents fill up their car with unleaded petrol. Perhaps they know about the 'ozone layer', 'acid rain', or 'the greenhouse effect'.

As teachers we should help children understand why it is important to look after the environment, and help them become aware of their responsibility to conserve resources and protect the world in which they live.

Children should investigate the extent to which everyday waste products such as paper, plastic materials and cans decay naturally. They should keep records of their observations and use this knowledge to help improve the appearance of their local environment.

Children should study aspects of their local environment which have been affected by human activity, for example, industry and mining. They should observe and record the significant features of the process.

ACTIVITIES

1. The big wrap!

Age range
Five to seven.
Group size
Individuals.
What you need
Packaging from sweets etc, scissors, pencils, paper, adhesive.
What to do
Ask the children to save all the wrappings from the sweets, chocolate bars, crisps and biscuits they have eaten during one week. How many wrappers are there on each packet?

The children should stick the wrappers on a large piece of paper. Do they think they need all this wrapping paper? Are any sweets unwrapped?
Further activity
Ask the children to collect packaging from one shopping trip. Compare the number of items bought with the amount of packaging collected.
AT5/1a

2. Waste not, want not

Age range
Seven to eleven.
Group size
Whole class or small groups.
What you need
Disposable gloves, a variety of waste materials, including different types of paper, plastic containers, glass bottles and jars, bottle tops, cans, a small magnet.
What to do
Wearing the gloves, the children should sort the waste materials into different groups, such as plastic, glass, paper, aluminium cans and other metals. They can test for aluminium in the cans using the magnet, which will attract cans *not* containing aluminium, and have no effect on cans which do contain aluminium. Tell them which materials can be recycled and which cannot. Ask the children where the different items can be taken for recycling. Can the children see why it is important to recycle waste materials?
Science content
Many types of material can be recycled, including paper, glass, some plastics and scrap metals. Throwing these things away is an unnecessary waste of both energy and materials.
Further activity
The children can make recycled paper and find out about local recycling groups and collection points, such as bottle banks and aluminium can collections. You could also find the local address for Friends of the Earth or similar organisations and ask a member to visit your school to talk about their work.
Safety
Care must be taken when handling waste materials. Children should always wear disposable gloves and be very careful in case there are sharp items.
AT5/1a, 3b, 4a

3. Litter survey 1

Age range
Seven to nine.

Group size
Small groups or whole class.

What you need
Disposable gloves, paper, pencils, pens, bags for collecting litter.

What to do
The children can draw a plan of the playground and/or field and mark the positions of litter bins. Wearing their disposable gloves before touching any litter, they should go outside after playtimes and mark on the plan the places where they find litter. They can make a colour code for different types of litter. You should ask them the following questions.
• Where was there most litter dropped?
• When was there most litter?
• What sort of litter was found?
• Are more litter bins needed?
• Should the litter bins have lids?

Science content
Human activities produce a wide range of waste products.

Further activity
The children could carry out the survey over several days to look for patterns. Is there more litter on windy days?

Empty the litter bins after one playtime and sort the litter into groups, for example, sweet papers, crisp packets, cans, etc.

Safety
Disposable gloves should be worn when handling any litter. Take great care when handling litter as there may be sharp items. Some local councils provide litter-collecting sticks.
AT5/1a

4. Litter survey 2

Age range
Seven to nine.

Group size
Small groups or whole class.

What you need
Disposable gloves, bags for collecting litter, paper, pens, pencils.

What to do
Choose an area near the school, such as a playground or shopping precinct, in which to carry out a litter survey. The children should find out the answers to the following questions:
• How many bins are there?
• Are there enough?
• Is there any litter on the ground?
• Where is there most litter?
• How many different types of litter are there?
• Did they see anybody dropping any litter?
The children can record the results on a plan of the area and colour-code the different types of litter.

Further activity
Sort the litter into groups. The children could suggest solutions to the litter problem, such as providing more litter bins, using different types of bins or positioning the bins differently.

Safety
If litter is to be handled, disposable gloves should be worn. Warn children to look out for sharp items. Use litter-collecting sticks if available.
AT5/1a, 3b

5. Going, going, gone!

Age range
Seven to eleven.
Group size
Small groups.
What you need
Disposable gloves, vegetable peelings, cubed bread, fruit and vegetables, five shallow dishes, food-wrapping film.
What to do
Wearing disposable gloves the children should put one piece of each of the foods into each of the five dishes.
• Dish 1: cover with film.
• Dish 2: add a few drops of water and cover with film.
• Dish 3: leave dry and uncovered.
• Dish 4: add a few drops of water.
• Dish 5: leave dry and uncovered and in a warm place, for instance, near a radiator.

Every day the children could record any changes they notice in the food in each dish.

Science content
Various factors affect the decaying process, including moisture, light, temperature, exposure to micro-organisms, and compactness.
Safety
Disposable gloves should be worn. Care must be taken in disposing of the decaying materials.
AT2/4b
AT5/2a

6. Leftovers

Age range
Five to nine.
Group size
Individuals.
What you need
Disposable gloves to handle litter.
What to do
The children can ask their parents to help them with this activity. They should make a list of everything their family throws away during one weekend. They can sort the items into groups of, for example, paper, cans, plastic and vegetable peelings.

Science content
Children should know that human activities produce a wide range of waste products.
Further activity
The children can carry out a survey of classroom litter which they collect from the classroom bin at the end of each day. They should weigh the litter collected each day and find out how much has been collected at the end of the week.
Safety
Disposable gloves should be worn to handle any litter, and hands should be washed thoroughly after this activity.
AT5/1a, 4a

7. Is there any pollution in the air?

Age range
Nine to eleven.
Group size
Small groups.
What you need
Double-sided sticky tape, yoghurt pots covered with white fabric secured with elastic bands, microscope or magnifier.
What to do
Set the children the following tasks:
• Put some double-sided sticky tape on several window ledges around the school. Inspect the sticky tape over a period of time. Does it get dirty?
• Put yoghurt pots covered with white fabric on the window ledges. Again, observe them over a period of time. Is there any dirt on the white fabric? Which of the window ledges are exposed to the most dirt? What do the children think might be the reason for this?
Science content
Air pollution is mainly caused

by domestic and industrial emissions of gases and solids. The distribution of the dirt caused by air pollution is affected by wind direction and the location of the sources of pollution. Common pollutants are sulphur dioxide, oxides of nitrogen, carbon monoxide, smoke and dust particles.

Further activity

With the children, collect some leaves from the roadside. Ask the children to put sticky tape on the surface of the leaves, peel it off and see if there is any dirt on it. This is more easily seen if the pieces of sticky tape are stuck on to white paper. They can also look at the tape under a magnifier or microscope.

Safety

When the children are collecting leaves near a road, *careful* supervision is necessary.

AT5/3a, 5a

8. Hunting for lichens

Age range

Nine to eleven.

Group size

Small groups.

What you need

Recording sheets, pencils.

What to do

The children should walk around the school grounds looking for lichens on paving stones, the tops of walls and the bark of trees. Are the lichens of the crusty, leafy or shrubby type? Which type of lichen did they find most often? Are there any types that they have not found at all? What does this tell them about the air pollution around the school?

Science content

Lichens are good indicators of air pollution in a locality. They are two plants, an alga and a fungus, which live together for mutual benefit. Some lichens are very sensitive to the amount of sulphur dioxide in the air.

• Crusty lichens look like grey-green crazy paving, and are found on walls, stones, roofs and trees. If the children find *only* crusty lichens it means that the air is fairly polluted.

• Leafy lichens look like flat rosettes of thin leaves, and are found on walls, stones, roofs and trees. They can stand a *little* air pollution.

• Shrubby lichens look like greenish-grey flattened branches, and are found on the branches and trunks of trees. These lichens will usually only grow where there is *no* air pollution.

• The children may not find any lichens at all, but they might find a bright green alga called Pleurococcus on the barks of trees. This may mean very bad air pollution.

AT5/3a, 5a

Crusty lichens

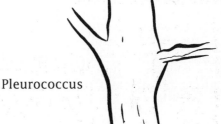

Pleurococcus

Shrubby lichens

Leafy lichens

9. The rot sets in!

Age range
Seven to eleven.
Group size
Small groups or the whole class.
What you need
Disposable gloves, a diary or recording sheet, a pencil, a trowel, various waste items like paper, vegetable peelings, apple core, banana skin, drinks cans, milk bottle tops, plastic bottles, polythene bags, flags for marking the position of litter.
What to do
Wearing the disposable gloves the children should bury some items of litter outside in the soil or in a basin of damp soil. They then mark the position of each item with a flag. What do they think will happen to each item after it is buried? They can look at the litter once a week for several months and see what changes (if any) have taken place. They should also write down the results in their diary.

Science content
Some items decay naturally. They are biodegradable – they can be broken down by biological means, by bacterial action. Plastics are made from oil, and are not biodegradable.
Further activity
Ask the children to investigate how uncovered litter decays. They could leave similar items of litter open to the air and secure them to the ground with some wire netting. They can look at the litter over a period of time and note any changes.
Safety
Disposable gloves should be worn when handling waste materials, and hands should be washed thoroughly afterwards.
AT5/2a, 2b, 5b

10. Turbulent water

Age range
Nine to eleven.
Group size
Small groups.
What you need
Water samples, measuring cylinder, white paper, black felt-tipped pen.
What to do
Ask the children to mark a black cross on a piece of white paper and put a measuring cylinder on top of the cross.

They can look down into the cylinder to see if they can see the cross. They should then pour water from one of their samples slowly into the cylinder until they can no longer see the cross, and read off the level of water in the cylinder. They should repeat this activity with their other water samples. Which water is the cloudiest?
Science content
The shorter the column of water is when the cross disappears, the cloudier the water. Do not assume that all dirty-looking water is polluted. It may be cloudy due to silt or sand deposits.
Further activity
Make a water filter using a one-litre plastic bottle and a funnel, some sand and some gravel and stones. The children could filter the water through layers of sand, gravel and stones in the funnel.
AT5/5a

11. Removing oil from troubled waters

Age range
Nine to eleven.
Group size
Small groups.
What you need
Salt water, a small quantity of crude oil, detergent, pieces of polystyrene, plaster of Paris, four small beakers, a dropping pipette, sawdust.
What to do
Ask the children to half fill each beaker with salt water. Using the dropping pipette, they can add some oil to each beaker, drop by drop, until the surface of the water is just covered. They can then add plaster of Paris to one beaker, some pieces of polystyrene to

another, a small quantity of detergent to the third beaker and sawdust to the fourth beaker. They should look at all the beakers after five minutes and record what has happened to the oil. Is the oil still on the surface of any of the beakers? Did the oil sink to the bottom of any of the beakers? Which method is most effective for removing oil from water?

Science content
Plaster of Paris will absorb the oil and sink to the bottom of the beaker. Oil will collect on the polystyrene and the sawdust at the surface of the water. Detergent disperses the oil throughout the water.

Further activity
Relate this work to oil spillages and polluted beaches.

AT5/3a, 5a

12. Bottleneck

Age range
Ten to eleven.
Group size
Whole class.
What you need
Photocopiable page 184.
What to do
Bottleneck is a small rural town on the busy A108 road to the coast. Not only is it busy in summer with holiday makers, but lorries use the road every day taking goods to and from Europe.

A recent survey has shown that many of the old buildings in the town are becoming damaged due to the heavy traffic passing through the town. The ancient parish church is in urgent need of restoration, and the house where the famous scientist Sir Harry James lived is in danger of collapse. The local archaeology society has been campaigning for a bypass for years. It is supported by local residents who dislike the noise and smell of cars and heavy lorries, and cannot shop in the summer due to the traffic jams.

The archaeology society has suggested two possible routes for the bypass. Another group of residents has suggested a third route.

The children should see whether they can find out:
• Who might be against having a bypass around Bottleneck? Explain why they would be against the bypass.
• Who might object to Route 1 of the bypass, and why?
• Who might object to Route 2 of the bypass, and why?
• Most people think that Route 3 is not feasible. Others think it is the best route. The children should try to explain these different views.

Further activity
Other examples that could be used for a role play:
• The pedestrianisation of a shopping street in the town.
• The Channel tunnel and associated road routes.
• British Coal wants to mine in the area.

AT5/5c

CHAPTER 8

Studying the environment

The environment has become a very prominent issue, often receiving prime-time media coverage. It is important that children become aware of the importance of the local and global environment, and that they learn to show animals and plants the respect which they deserve. This is an excellent area for enjoyment and for study, and one which need not involve much expensive equipment.

Children should develop their knowledge and understanding of the variety of animals and plants, the ways in which they are classified and the localities in which they live. By caring for animals and plants in the classroom, children will learn to treat living things with care and consideration and to understand their needs for a healthy life.

The fantastic variety of plants and animals may have to be studied through secondary sources in part, but do not forget the wealth of material which is beneath every stone, in leaf litter, on grasses and shrubs and in ponds. What excitement and noise as we found a great diving beetle and then a water scorpion – and this was on a course for teachers!

As their understanding develops, children become more sophisticated at classifying and identifying plants and animals using keys. They may study different localities at various times of the year, and will begin to appreciate why different organisms live in different places. At the same time, they will develop their ideas about predators and prey. All in all, the topic of the environment provides a feast of interest.

ACTIVITIES

1. Variety is the spice of life

Age range
Five to seven.

Group size
The whole class.

What you need
Pond nets, shallow white dishes, equipment for collecting and observing animals, including hand lenses, hand magnifiers, pooters, plastic teaspoons, plastic collecting jars.

What to do
Ask the children to stand at the edge of the pond and put some pond water into a shallow dish. Without stirring the mud they should move the net to and fro and up and down in the pond. Then they can turn out the contents of the net into the dish of water. The children should observe any pond animals and plants. Are they round, long, large, small, hairy? Do the plants have large, flat leaves or thin, feather-like leaves? Encourage the children to use words such as snails, worms, insects and beetles when describing the creatures. Make sure that they do not forget to look at the plants, noticing, for example, whether they are partially immersed in the water. The children could then:
• draw the animals and plants with the aid of a hand lens or magnifier;
• describe the animals and plants to their friends;
• try to identify an organism using a simple 'key' (a small book or chart);
• use books, stories, pictures, charts and videos to find out about a variety of animal and plant life.

Science content
There is a wide variety of living things, which can be divided into two groups, animals and plants. Insects are animals, as are dogs, cows, birds and humans.

Further activity
The children can carry out other surveys of different life forms. They could try some of the following activities.
• Shake the branches of a tree over a white cloth and collect and examine the creatures which will fall out, using pooters and teaspoons.
• Sweep a large pond net through some long grass, place specimens in white trays and examine them.
• Carry out a minibeast hunt in the school grounds using pooters, teaspoons and collecting jars.
• Visit a zoo and see all the different types of animals.
• Collect books, charts and videos which illustrate the variety of animals in the world.

Safety
Careful supervision at water locations is essential. Steep-sided, deep ponds and canals should be avoided, particularly with younger children. Plastic containers are preferable to glass ones.

AT2/1a

2. Caring for classroom animals

Age range
Five to seven.

Group size
Small groups or the whole class.

What you need
A small animal such as a hamster, gerbil or guinea pig, suitable housing, variety of bedding, variety of *safe* foods, objects for the animal's living quarters such as balls, sticks and blocks.

What to do
Discuss with the children their own needs, such as shelter, food, water, clothing, safety, security, toys and friends. Relate these needs to those of other animals.

Set up the cage or hutch, making sure that any items you put inside are safe, and then introduce a class pet.

The children can take it in turns to feed, water and clean out the animal. It is important to organise what happens when someone is absent, and to work out who will care for the animal at weekends and during the holidays. Some animals may be safely left over the weekend in cool parts of the classroom (avoid direct sunlight) but they cannot be left for long periods of time. Arrangements could be made to send them on their holidays to 'good' homes – or be prepared to take them yourself.

Science content
Animals in cages are totally reliant on their captors. They need shelter, food, water, the correct temperature and items of interest in their environment.

Further activity
Children could investigate whether the animal prefers one type of food to another. Do not use human food for this activity, and ensure that the food used is safe. Two or three trays of food should be put at one end of the cage, and then the animal can be left to choose. Ensure that a fair test is carried out by using equal amounts of food on similar trays. Show the children some books and leaflets on how to look after pets.

Safety
Children should always wash their hands after touching animals or animal bedding. Check to see whether any children have allergies to furry animals, dust etc.

A school policy on scratches and bites should be carefully considered.

AT2/2a, 2b

3. Similar and different

Age range
Seven to nine.
Group size
Small groups.
What you need
Animals collected from ponds, trees, grass and bushes, or from a minibeast hunt (see Activity 1, page 86).
What to do
Observation is a skill which needs to be taught, rehearsed and refined. The children should try to make detailed and accurate observations of the animals they have collected by describing them and by drawing and labelling pictures.

They can then try to identify the animals on the basis of their observable structural features, using simple keys or pictures.

Next the children can sort the animals into broad groups according to their observable features. First they can use their own groupings, such as length of legs, body shape etc, then they can find out about more conventional groupings, such as animals with six legs are insects, animals with eight legs are spiders (or arachnids), animals with more than eight legs are crustacea etc.

Science content
Living things show differences and similarities which form the basis of classification. Children initially group organisms according to their own criteria, which may not resemble accepted classification schemes. While this should be encouraged, children should gradually be introduced to the more usual classification schemes.
Further activity
By visiting one location at different times of day and throughout the year, children will begin to realise that living things respond to daily and seasonal changes. For instance, tadpoles are only found in the spring, butterflies only fly on warm, sunny days, and human beings are found lying out on the grass on hot days!

Encourage the more able children to identify the similarities and differences between a water louse and a freshwater shrimp (both crustaceans), or a water louse and a dragonfly larva.
Safety
Respect for living things should be encouraged at all times, and living things should always be returned to their environment, even if they are temporarily kept in school.
AT2/3a, 3b, 4a

4. Fruits and seeds, pips and stones

Age range
Five to seven.
Group size
Small groups.
What you need
A selection of fruit, for example banana, tomato, bean pod, walnut in shell, orange, apple, mango, peach, chilli, and wild fruit and berries such as hazelnut, crab-apple, hawthorn berry etc.
What to do
Ask the children to examine every aspect of the fruit.
• Let them describe the colours of the fruit, and put them into sets on the basis of colour. Are some difficult to place? For example, an apple might be yellowy-green.
• They should describe the shape of the fruit. How are they different and how are they similar? How do fruit of one colour differ in shape?
• Ask the children to describe the smell of the fruit. They must take care with hedgerow fruit. Are any of the smells similar? Do the fruit smell nice? Do all of the children like the smell of the fruit?
• Ask the children to describe the texture of the fruit. Are they rough, smooth, hairy or spiky?

Common pondsnail

Frog tadpole

Mayfly larva

Water louse

• Let them cut open the fruit with a blunt knife. What do the insides look like? Do they differ? The children should carefully observe the pips and stones. How are they similar or different? Do big fruit have more seeds than small fruit? Are their seeds bigger or smaller?

• Ask the children to describe the taste of the fruit. Again, care must be taken with hedgerow fruits. Are they sweet or sour? Do all the children like the same fruit?

The children can collate their observations on a database.

Science content

A fruit is the ripened ovary (female part) of the flower, and it contains seeds. We commonly use the word fruit to describe fleshy foods which are associated with desserts, but nuts and pea pods are also fruit as they contain seeds.

Children could look at the plants in and around the school, and identify their fruit. Berberis and cotoneaster at least will probably be present in the grounds of most schools. The children will use a number of skills in communicating and recording the results of their explorations, and in assessing the similarities and differences between plants.

Further activity

The seeds from the fruit could be saved and placed in compost or soil. How many of them grow? How long do they take to germinate? Do larger seeds germinate first? What conditions do the seeds need to grow? Encourage the children to grow seeds at home (see Activity 12, page 96).

The children could try rolling the seeds and fruit down a slope to see the effect of shape on movement.

Safety

Children must be warned of the dangers of tasting some hedgerow fruit. Stress that some plants, such as deadly nightshade, are extremely poisonous, and avoid letting the children taste anything unless you are entirely sure that it is safe. If the children are going to taste some of the fruit, knives and surfaces must be clean and the fruit should be peeled or washed.

AT2/1a, 3a, 3b
AT12/2b, 3b, 3c

5. Looking at rocks

Age range
Six to ten.
Group size
Small groups.
What you need
Rocks, hand lens, scales, coin, nail, vinegar, hammer and cloth bag, safety glasses.
What to do
Ask the children to find out as much as possible about a rock of their choice, perhaps using the chart below to give them ideas for their investigation. They can produce a poster with a picture of the rock and details of their findings.
Science content
The above exercise will give the children practice in:
• investigating a natural material using a variety of exploration skills;
• observing familiar materials at first hand using their senses;
• describing and communicating observations;
• identifying simple differences, such as rough and smooth;
• listing and collating observations;
• recording findings in charts, drawings and other appropriate forms;
• selecting and using simple instruments to enhance their observations.
Further activity
Ask the children to compare two different rocks, using criteria similar to the ones they explored with their original rocks.

A similar exercise could be carried out with soils, involving touching, observing, sieving and separating the soil. Is there any organic material in the soil?
Safety
If the rock is to be broken it should be placed inside a cloth bag in a safe location, and hit with a hammer. Safety glasses should be worn as an extra precaution.
AT9/3d

6. Air is all around us

Age range
Seven to eleven.
Group size
Pairs.
What you need
Large pieces of card, ruler, nail, two balloons, string, sticky tape, wooden stand made from a piece of dowelling with a block of wood as its base, Plasticine.
What to do
Ask the children to run a short distance in the playground, and then come inside and run the same distance in the hall or corridor (if this is allowed). They can then repeat this, but this time they should hold a large piece of card in front of them. What do they notice? Did it slow them down? Why is this happening? Ask the children if they can see a reason why cars don't have large flat fronts, but sloping, streamlined ones.

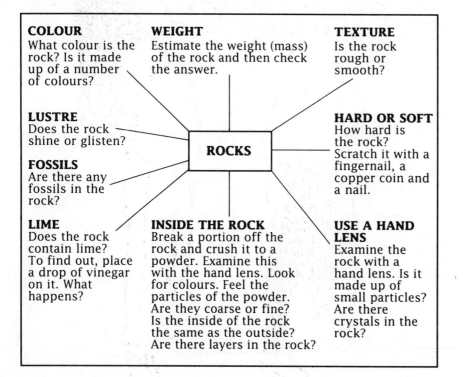

COLOUR
What colour is the rock? Is it made up of a number of colours?

WEIGHT
Estimate the weight (mass) of the rock and then check the answer.

TEXTURE
Is the rock rough or smooth?

LUSTRE
Does the rock shine or glisten?

FOSSILS
Are there any fossils in the rock?

LIME
Does the rock contain lime? To find out, place a drop of vinegar on it. What happens?

ROCKS

HARD OR SOFT
How hard is the rock? Scratch it with a fingernail, a copper coin and a nail.

INSIDE THE ROCK
Break a portion off the rock and crush it to a powder. Examine this with the hand lens. Look for colours. Feel the particles of the powder. Are they coarse or fine? Is the inside of the rock the same as the outside? Are there layers in the rock?

USE A HAND LENS
Examine the rock with a hand lens. Is it made up of small particles? Are there crystals in the rock?

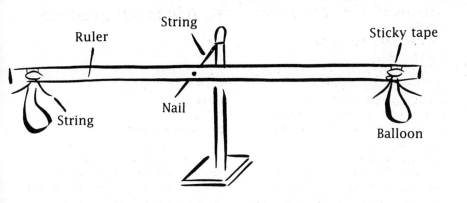

Ruler · String · Nail · String · Sticky tape · Balloon

Wind speed

One child in each pair should move the piece of card backwards and forwards like a fan. What does the other child feel? Is it like the feeling of being outside on some days?

The children can now set up the model illustrated above. They should attach the ruler to the wooden stand by fixing a nail through the centre of the ruler and attaching it to the post with string. Then they can use sticky tape to fix a deflated balloon to each end of the ruler, tying the balloons loosely with string. Ask them to make the ruler balance, if necessary, by weighting the ends with Plasticine.

Next they should blow up and tie one of the balloons without removing it from the apparatus. What happens? Can they see a reason for this?

Science content
• Air is all around us, and it resists our movement through it. At higher altitudes the air is thinner, and therefore there is less resistance.
• Wind is air in motion.
• Gases such as air have 'weight', or mass. This is demonstrated by the fact that the balloon filled with air weighs more than the balloon without air.
• Air is made up of many gases, such as nitrogen (78%), oxygen (21%) and carbon

dioxide (0.04%). It also contains water vapour, and often carries pollutants as well.

Further activity
Look at wind speeds in weather forecasts on the television or in newspapers.
AT9/3b, 4a
AT6/5a

91

7. Spot the difference

Age range
Nine to eleven.
Group size
Pairs or small groups.
What you need
Pond animals or plants, or other animals collected in Activity 1 on page 86, white trays, teaspoons, hand lenses or magnifiers.
What to do
Present the children with two animals from the same classification group, such as a water louse and a freshwater shrimp, which are both crustaceans. The children should then observe the animals and record any similarities and differences they can find, using annotated diagrams or tables.

Next let the children examine two animals from different classification groups, such as a water louse (crustacean) and a dragonfly larva (insect). Again, they should observe the animals and record any similarities and differences using annotated diagrams or tables.

Science content
This activity gives children the opportunity to identify similarities and differences, both within a group of animals or plants and between groups of animals or plants. Through activities like this, children should begin to see and understand the principles on which classification systems are built.
Further activity
Repeat the activity, but use different animals which you may have collected from trees, grass sweeps or the minibeast hunt.
Safety
Ensure that children do not mistreat animals by rough handling or by taking aquatic animals out of the water. Always try to return animals to their natural environment after study.
AT2/4a

8. Fossils

Age range
Nine to eleven.
Group size
Pairs.
What you need
Fossil specimens, charts, books or videos about fossils, Plasticine, coins, shells, keys and other small objects, petroleum jelly, plaster of Paris, paper or plastic cups.
What to do
Present the children with a number of fossils, pointing out if the material fossilised is rock, carbon film, bone or shell etc. Ask them to discuss the fossils, and suggest how they might have been formed and what the fossilised object might originally have looked like.

|——| Actual size

Freshwater shrimp

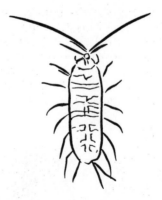

Actual size

Water louse

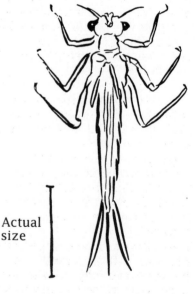

Actual size

Dragonfly larva

The children can then try making their own 'fossils'. They should first coat the outside of the object to be cast with petroleum jelly. Then they should press the object firmly into the Plasticine to obtain a complete impression and then remove it. Make a smooth mixture of plaster of Paris and water, and make sure the children build up the edges of the cast with Plasticine before pouring in the plaster of Paris . The casts should be left to set, and then the children can carefully remove the mould and examine their 'fossil'.

Science content

The study of fossils is called palaeontology. A fossil is the remains of an organism, or direct evidence of its presence, preserved in rock. Generally only the hard parts of the original animal or plant are preserved and are usually partly or wholly replaced by minerals in the ground. Fossils give us information about the past. They can tell us about the evolution of the horse over 60 million years, or that the coelacanth has hardly changed in millions of years.

Through handling fossils or fossil casts, children will begin to see that plants and animals can be fossilised in many ways.

Further activity

Consider life forms which became extinct a long time ago, like dinosaurs, and those which died out more recently, like the Tasmanian Tiger.

Safety

If fossils are to be chipped out of rocks, safety glasses must be worn. You may need to get permission to collect fossils in some areas, and wherever you collect them you should do so in a sensitive manner.

AT2/4c
AT4/3a

9. How tall is that tree?

Age range
Nine to eleven.
Group size
Pairs.
What you need
Metre rule, ruler.
What to do
Ask the children to use one or more of the methods described below to find the height of a tree or building.

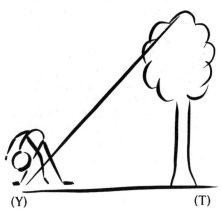

(A)

(B)

(Y)   (T)

• Method 1: the children bend down and look through their legs at the tree, and move until they can just see the top of the tree. The distance from them (Y) to the tree (T) should be similar to the tree's height.

• Method 2: ask one of the children to stand beside a metre rule at the side of the tree. Another should then hold out a ruler, and adjust it so that the top of the ruler is at the same level as the top of the metre rule. Then she should move her thumb to the point on the ruler which corresponds to the base of the metre rule. The children can then calculate how many times the section of the ruler AB which is the equivalent of one metre goes into the height of the tree.

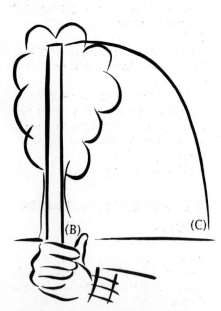

• Method 3: one of the children should stand at a distance from the tree and hold her arm out straight, holding a ruler or stick. She should then move her arm until the top of the ruler is at the top of the tree, and her thumb is at the base. Next she can rotate the stick through 90°, ensuring that her thumb stays at the base of the tree. Another child should then stand at the position indicated by the end of the ruler (C). The children can then measure the distance from the base of the tree to position C, to find the height of the tree.

The children can then use the three methods to measure the heights of some other landmarks, and record their results in a table like the one shown below.

Science content
The children have a number of methods by which to estimate heights accurately. These methods should be related to mathematical work. For example, in Method 1 the distance YT is equal to the height of the tree because the three points form an isosceles triangle.

Further activity
Are the results gained by each method similar? Which method do the children think is the most accurate, and which the least? See whether the children can devise a method to check this.

AT4/4a

Approximate height of	Method 1	2	3
Tree			
Church steeple			
School roof			

10. What conditions do woodlice prefer?

Age range
Ten to eleven.
Group size
Small group.
What you need
A range of equipment, including plastic containers or tanks, black paper, lamps and torches, stones, bark, soil, sand, cotton wool, sticky tape, stop-clock, water, plastic spoons, woodlice.
What to do
The children could go on a woodlice hunt, recording places where they find them. Discussion will reveal that woodlice are most often found in damp places, under stones, rotting wood and other objects. Ask the children why

they think this might be and help them to formulate a hypothesis. Do they think woodlice live where they do because they like damp conditions, or because they like the dark – or do they need both darkness and damp?

Next ask the children to design and carry out an investigation to test their hypothesis. They should record accurately the steps of the investigation, and their reasons for taking them. Encourage the children to work systematically, and ask them to record their findings in an appropriate way. How will they know when their hypothesis has been verified or otherwise? Was their test a fair one?

One good way of carrying out the investigation would be to use a tray lined with black paper and divided into three sections – one dark and dry (covered with black paper), one dark and damp, and one light and damp. No other materials need be used, as they increase the number of variables.

Ask the children to decide how many woodlice they will

use (a minimum of ten), the number of readings they will take and at what intervals. Children will often only take a reading when it fits in with their hypothesis.

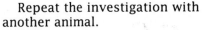

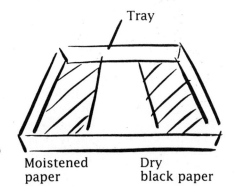

Moistened paper   Dry black paper

Science content
Woodlice have become adapted to a terrestrial life and are able to breath moist air through gills on the underside of their bodies. Woodlice belong to the class *crustacea*. They have seven pairs of jointed legs and feed on decaying vegetation, seedlings and ripe fruit.

Further activity
The children can find out more information about woodlice.

Ask the children to discuss why many people dislike 'creepy-crawlies' but like furry animals.

Repeat the investigation with another animal.
Safety
Hands should be washed after handling animals, soil or compost.

Stress the need to care for animals. Use plastic spoons to pick up the woodlice, as these should cause them no harm if used carefully.
AT2/2a, 2b, 5a

11. Birds' favourite foods

Age range
Five to nine (first activity) and ten to eleven (second activity).
Group size
Small groups.
What you need
Variety of bird seed sorted into different seed types, bread, pieces of apple, potato etc, pieces of bread coloured with different food dyes.
What to do
This activity is concerned with birds' food preferences. Ask the children to sort the seeds into sets, identifying the similarities and differences. They can then discuss the best place to put food out for the birds, perhaps on a bird table or tray. The children should decide how many pieces of each food are to be put out. Do they think this is important? Should the pieces of food be the same size? Discuss the idea of a fair test; younger and less able children will probably not see the need for it.

The children can then put out the food, and record the order in which the different foods are eaten. Is a particular food eaten by a particular bird? Discuss the feeding preferences of different types of birds; for example, some eat seeds and others do not.

For the second activity the children should be asked to design an investigation to see which colour of bread is preferred by a particular species of bird. Encourage them to design a fair test and to record their findings in an appropriate manner. For example, they could produce a report for the local ornithological society.

Science content
Some birds, like greenfinches, are seed eaters, others, such as kestrels, are meat eaters, while others, such as blackbirds, are mainly carnivorous, but will generally be tempted by bread, particularly in winter. Do not feed birds during the breeding season, as they should rear their young on natural foods.

The design of 'fair tests' should be encouraged from an early age, even if it takes a long time for the children to attain any degree of sophistication. In the second test above, it is important that the pieces of bread should be the same size, and that they all either have or do not have a crust. The same number of drops of colouring should be added to the bread in a constant volume of water, so that the intensity of colour is similar – but this presumes that the original colouring of the food dye was of the same intensity, which may not be the case. This shows how hard it is to make a test really fair.

Some children may realise that the positioning of the food is important. For example, they should ensure that all the green pieces are not nearest the direction from which the birds approach. A suggested arrangement is shown below.

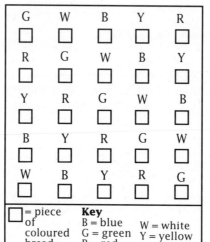

G	W	B	Y	R
☐	☐	☐	☐	☐
R	G	W	B	Y
☐	☐	☐	☐	☐
Y	R	G	W	B
☐	☐	☐	☐	☐
B	Y	R	G	W
☐	☐	☐	☐	☐
W	B	Y	R	G
☐	☐	☐	☐	☐

☐ = piece of coloured bread

Key
B = blue   W = white
G = green   Y = yellow
R = red

Further activity
• What food does the guinea pig, hamster or mouse prefer?
• See whether the children can find out more about the feeding habits of birds.
• Encourage the school or individual children to join 'Watch'/RSPB.

Safety
Use food colouring sparingly. Do not let children feed pets with dangerous substances or ones which might upset them. Use small amounts. If in doubt, leave it out!

AT2/4a

12. Which grows best?

Age range
Five to eight (first activity) and nine to eleven (second activity).

Group size
Small groups.

What you need
Yoghurt cartons or plant pots, soil or compost, seeds such as cress, french marigold or peas.

What to do
For the first activity ask the children to plant the seeds in four containers filled with soil:
• one with no water;
• one with no light (put in a cupboard or cover with black paper);
• one with no warmth (put in a fridge);
• one with water, light and warmth (in the classroom).

Ask the children to predict in which conditions the seeds will germinate and grow quickest and 'best'. What does 'best' mean? Which seeds germinated? Which plants look healthy? Can the children now tell which conditions seeds and plants need to germinate and grow?

The children should record their findings in pictures or words.

For the second activity ask the children to plan and carry out an investigation to see if they can find out what is the optimum amount of water or light and warmth for the successful germination of seeds and growth of plants.

Ask the children to record their findings in a report, including tables and graphs.

Science content
Most seeds require water and warmth in order to germinate. Light is generally not needed for germination, but some seeds do require it. The best conditions for healthy growth will generally include light, water and warmth.

These activities are particularly designed to develop exploration skills such as hypothesising and predicting, designing and carrying out investigations, interpreting findings, drawing inferences and communicating investigations and findings.

Emphasise the importance of trying to make investigations fair. In the first activity, similar containers should be used, with the same type and quantity of compost or soil. All the conditions (variables) should be the same except the one actually under consideration.

Further activity
Try using different concentrations of fertilizer, to see how this affects plant growth.

Safety
Children should wash their hands after using soils, compost and fertilizers.

AT2/2a

13. What shall I have for my breakfast today?

Age range
Ten to eleven.

Group size
Small groups.

What you need
Photocopiable page 185, scissors.

What to do
After visiting a tree, and possibly collecting specimens by shaking a branch on to a tray or white sheet (see Activity 1, page 86), ask the children how different animals get their food. What eats what?

Cut out the 20 pictures from photocopiable page 185 and place them at the appropriate place on a large diagram of a tree. See whether the children can gradually assemble a food web. You can then discuss with the children the number of organisms and/or the amount of energy at each level of the food web.

Science content
Plants are producers of food and energy. Using energy from the sunlight, the green pigment chlorophyll in the leaves and stems of plants uses the process of photosynthesis to produce glucose by combining carbon dioxide and water. The glucose is stored as starch, which gives energy to animals who eat the plants.

Herbivores or primary consumers are animals which eat plants. Carnivores are animals which eat other animals. Secondary carnivores are carnivores which eat other carnivores. Decomposers are organisms which break down dead plants and animals.

The oak tree is visited by a number of animals, both predators and prey. Oak leaves make their own food; caterpillars eat oak leaves; warblers eat caterpillars; sparrow-hawks eat warblers. This makes up a food or energy chain (where ↓ means 'eaten by'):

Oak leaf

↓

Caterpillar

↓

Warbler

↓

Sparrow-hawk

However, oak leaves are not eaten by caterpillars alone, and sparrow-hawks eat other things as well as warblers. A food or energy web is made up of many chains like this one, and the combination of all these chains can give us a more complete picture of the feeding or energy dynamics of an environment.

Further activity
A similar exercise could be carried out with pond plants and animals.

AT2/5d

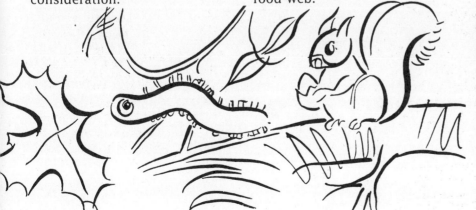

CHAPTER 9

Water

The sea is the major source of our drinking water, yet unless we use expensive desalination plants, we cannot drink it, at least not as it is! This chapter should help older children to explain the water cycle and to understand the concepts of evaporation, cloud formation, cooling as the water vapour rises and precipitation. Meanwhile, younger children will be able to experience playing with water and will discover sinking and floating objects.

Water is vital to life; about four fifths of our bodies are water. Children can investigate the water content of foods, and see which substances dissolve in water and which do not, as they start on the long road to understanding the idea of particles and atoms.

Water is a cause of rusting, which gives rise to a lot of expensive damage, so it is useful for children to look at factors which speed up and prevent rusting.

The shape and speed of boats moving through water, water pressure, and the tendency of water to reach its own level can all be investigated, as the classroom floor gets wetter and wetter!

The theme of water allows investigations using harmless chemicals which are both exciting and colourful. Children can watch as crystals grow, and can determine whether everyday substances such as sugar, salt and vinegar are acid, alkaline or neutral by using red cabbage water as an indicator – just watch the colours change!

Finally, the concept of separation may be developed by filtering salt and sand, and by watching the movement of colours on filter or blotting paper.

ACTIVITIES

1. Water content

Age range
Six to ten.
Group size
Small groups.
What you need
Bread, cucumber, biscuit, carrot, grass etc.
What to do
Ask the children to weigh out small amounts of common foods and commonly-occurring natural substances. It is a good idea to use samples which have the same weight, to make comparisons easier.

Item	Appearance at start	Mass at start	Mass at end	Change in mass	Appearance at end
Bread	White, soft				

Next, the children should place the items on plates or tins and put them on a radiator, next to a sunny window, or in an oven at 60°C. They should make sure that none of the items burn. Once the items are dry they can be weighed again and the results recorded in a table.

Science content
All living things contain water. Water accounts for about four fifths of our bodies by weight, and some vegetables can be 90 per cent water or even more. Contrastingly, cereal grains contain as little as 14 per cent water.

Water is removed by evaporation and heating, but some substances such as biscuits absorb moisture from the air, and therefore may gain weight at room temperature.

Safety
Care is obviously required when using ovens. Food items should not be eaten if they have been cut with dirty instruments, or if they have been handled or left in the classroom for a while.
AT2/3a

2. Shape, speed and ships

Age range
Nine to eleven.

Group size
Small groups.

What you need
Guttering, clothes pegs, wooden dowels, two cotton reels, length of cotton, Plasticine, support to hold upper end of pulley, three different boat shapes.

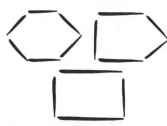

What to do
Set up the equipment as shown in the diagram. Use as high a table or bench as possible, as the distance from the top cotton reel to the floor will be equal to the distance travelled by the boat.

Attach the end of the cotton to each boat in turn, and ask the children to investigate the time it takes each boat to complete one length of the guttering. They should adjust the Plasticine mass so that the boats will just move. The children will then be able to record their results in a table like the one below.

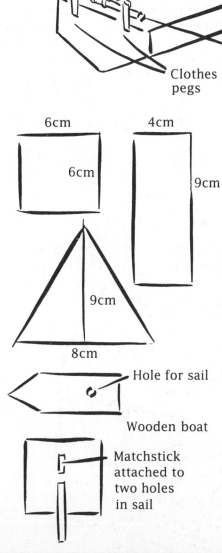

Support B
Plasticine mass/weight
Cotton
Boat A
Guttering containing water
Dowel (wood)
Clothes pegs

6cm
6cm
4cm
9cm
9cm
8cm
Hole for sail
Wooden boat
Matchstick attached to two holes in sail

Science content
The shape of the boat affects its speed through water. The less resistance provided by the boat's contact with the water, the faster the boat will go.

Further activity
Ask the children to investigate how the shape of a sail affects the speed of movement. They should use the fastest boat shape from the above exercise and try sails of various shapes; square, oblong and triangular. These should be designed to have the same surface area, to ensure a fair test.

The children could also try propelling the boat by means of a battery-operated hand-held fan, held at a fixed point.

Safety
Take care always to keep water away from electrical apparatus and sockets.

AT10/3a

Boat	Time 1	Time 2	Time 3	Average time or median time

Substance	Appearance	Does it dissolve?	Appearance after
Coffee	Brown granules	Yes, quickly	Brown liquid
Sugar			

3. Will it dissolve?

Age range
Six to ten.
Group size
Small groups.
What you need
Clear plastic containers (such as jam-jars), coffee, sugar, salt, flour.
What to do
Ask the children to part-fill some containers with water and then predict what will happen if they sprinkle the various substances on the water. They can then do this, carefully watching what happens. Does the colour of the water change? They can fill in a table like the one shown above.

Science content
Some substances can be dissolved in water, others cannot. When a solid dissolves it is still in the water, even if it cannot be seen. It fills the gaps between the water molecules. You can prove that a dissolved solid is present in solution by heating the mixture – this will return it to its solid form.

Substances dissolve faster if they are stirred, if the liquid is warm or if they are in powder form, dispersed through the liquid. A powder floating on top of a liquid may dissolve more slowly than a solid of the same type and amount, as little of it is in contact with the water. A solid dissolves more rapidly in warm than in cold liquid. When a solid dissolves in a liquid a *solution* is produced.

Further activity
The children can try to find out how stirring or heating the water affects the speed at which a substance dissolves. Does a whole sugar lump or a crushed sugar lump dissolve quicker? Does a whole indigestion tablet dissolve faster than a crushed one?

Safety
Care needs to be taken if glass jam-jars are used.

AT6/2b, 3b

4. Stop rust!

Age range
Seven to eleven.
Group size
Small groups.
What you need
Dry, clean, shiny nails, jam-jars, paint, grease such as petroleum jelly, cooking oil.
What to do
Ask the children to part-fill four jam-jars with water and put a nail in each. One nail should first be coated with paint, another with grease and another with cooking oil while the fourth is left as it is.

The children should leave the nails in the water for about a week and then inspect them for rust. What are the results? How do they interpret these findings?
Science content
Both iron and steel may rust. Rust is a brownish-red coating which forms on the surface of these two metals when they are unprotected and in contact with moisture and oxygen from the air. However, the metals can be prevented from rusting by covering them with oil, paint, grease or non-rusting metals, such as chromium.
Further activity
The children could look for signs of rust, for example on cars and railings. They could make a list of iron and steel objects and the ways they can be protected from rusting, for example chromium-plated car bumpers or painted railings.
Safety
Take care when handling rusty metal.
AT6/4a, 4b

5. Soaking up

Age range
Five to seven (first activity) and nine to eleven (second activity).
Group size
Small groups.
What you need
Pieces of fabric, water, jug or measuring cylinder, other equipment such as scales and forceps.

What to do
The first activity could be put in the context of a water spillage in the classroom or kitchen, or Teddy needing suitable clothes to wear while walking in the mountains.

Ask the children to touch the fabric pieces and describe the differences and similarities between them. For instance, do they differ in colour and pattern or texture and size? This helps to start to develop the concept of a fair test.

Ask the children to predict which material would be best for mopping up a spillage of water. Can they think of a way of finding out? They should try out their ideas and record their findings in a suitable way, perhaps on a poster.

For the older children, ask them to use any available equipment to design an investigation to see which fabric absorbs most water. Tell them to make sure they have a fair test. They should record the results in a suitable format, such as a poster or table.
Science content
Some fabrics, such as wool, absorb large amounts of water while others do not. Some fabrics have water-repellent chemicals sprayed on to them.

The emphasis of this activity is on devising a fair test by such means as using equal sized pieces of fabric, and keeping them immersed in the same amount of water for the same time. Other exploration skills will also be highlighted when the children ask questions such as 'What would happen if we...?'. The activity also gives a chance to develop children's recording skills.
Further activity
Ask the children to look at fabrics under the microscope.
AT6/4a, 4b

6. Water pressure

Age range
Nine to eleven.
Group size
Small groups.
What you need
Plastic container, for example a washing-up liquid bottle, with four evenly spaced holes along its length, sticky tape, sink or bowl, food colouring.
What to do
Ask the children to predict from which hole the water will shoot the furthest if the bottle is filled with water. They should then stick tape over the holes and fill the bottle with water. They can use food colouring for added effect. Over a sink or bowl, they can remove the tape while watching carefully. Ask them to record what they see in the form of a diagram.

Science content
Water exerts a pressure on objects with which it comes into contact. This pressure increases with depth, which is why deep sea divers wear strong diving suits to resist water pressure. Likewise, dams and submarines are made of strong, thick materials to withstand the immense pressure exerted on them by water.
Further activity
Ask the children to investigate the time it takes for 100cm³, 200cm³ and 400cm³ of water to escape from a hole at the base of a plastic container. They should first try to predict the outcome, then investigate whether this is so, record their results and explain their findings.
Safety
Ensure that the tape is removed a safe distance from electrical items and sockets.

7. Water levels

Age range
Eight to eleven.
Group size
Small groups.
What you need
Two metres of plastic tubing, two plastic funnels, water, food colouring.
What to do
Ask the children to fix one funnel into each end of the clean plastic tubing. Make sure that they fix the funnels in securely, as this will help to reduce the risk of spillage whilst pouring. The children can then pour in some water mixed with food dye until the tube is two thirds full.

Water at same level in each tube

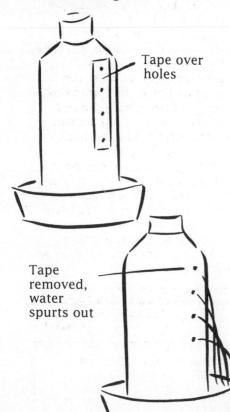

Tape over holes

Tape removed, water spurts out

Ask the children to predict what will happen if one or other funnel is raised or lowered. What do they notice? What always happens? You may find that a sheet of horizontally lined paper on the wall will help to show what is happening.

Science content
The water will always reach the same level in each side of the tubing. Water always finds its own level. In the diagram below, for example, water reaches houses from a distant reservoir because it flows downhill. Hill 'W' does not prevent its passage because water is capable of reaching its own level. Therefore, houses A and B will get water but house C will not, unless a pump is used, because it is higher than the reservoir.

Further activity
Investigate with the children the way that water drains off buildings.

Safety
Ensure that water is kept well away from all electrical appliances and sockets.

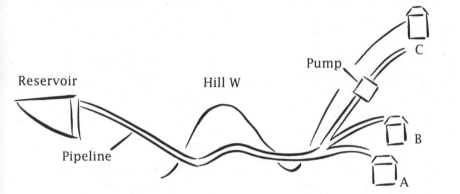

8. The acid test

Age range
Ten to eleven.

Group size
Small groups.

What you need
Red cabbage (fresh or frozen), test tubes or yoghurt pots, hot water in a container, scissors or knife, dropping pipettes, everyday substances such as vinegar, soda, tea, coffee, lemon juice, lemonade, salt, sugar, milk, bath crystals, soap, bicarbonate of soda (some of these will need to be dissolved in water).

What to do
Ask the children to cut up the red cabbage, place it in hot water and squeeze the colour out with a spoon. The water should turn a light purple colour. Direct heat is not really needed unless the cabbage is dry. The children can then place a few drops of the cabbage water in each test tube or yoghurt pot, using a dropping pipette. They can then mix one or two drops of each substance with the red cabbage water, using a clean pipette each time, and shake the containers gently.

What do the children notice when they mix the red cabbage water with each substance?

The results can be recorded in a table and the children can use them to find out which substances are acid, neutral or alkaline.

Substance	Colour with red cabbage	Acid, alkali or neutral
Vinegar	Red	Acid

Science content

Certain substances such as litmus can *indicate* whether various solutions are acid, alkaline or neutral. We call these substances indicators. Red cabbage water is a good indicator as it distinguishes between acids, alkalis and neutral substances. If the cabbage water turns red the solution is acid; if it turns green the solution is alkaline, and if the water stays light purple the solution is neutral.

It is important to ensure that the test is fair. Equal quantities of each substance should be tested with equal quantities of red cabbage water, bearing in mind that it is difficult to standardise the strength of each substance.

Further activity

Ask the children to use universal indicator solution or universal indicator paper and test the substances again. Universal indicator indicates the degree of acidity or alkalinity of a liquid as measured by pH.

Use some other substances, such as berries, and see whether they are good indicators.

Safety

Care should be taken when selecting the substances to be tested. Avoid using dangerous liquids such as bleach.

AT6/5b

9. Growing crystals

Age range
Seven to eleven.
Group size
Small groups.
What you need
Epsom salts or alum, plastic beaker or yoghurt carton, microscope and slide, plastic teaspoon, dropping pipette.
What to do
Ask the children to put about 10cm³ of warm water into a beaker or yoghurt pot and add the Epsom salts or alum, stirring with a plastic spoon until no more will dissolve. A solution in which no more solid will dissolve is called a saturated solution.

The children can then put a drop or two of the solution on a microscope slide and watch through the microscope as the liquid cools and evaporates.

They can then draw the crystals which appear, and describe the shapes of the crystals.

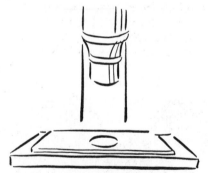

Crystals form as
liquid evaporates

Science content
Crystals are solids with regular shapes, flat surfaces and straight edges. They are present in rocks. Everyday items such as sugar granules, salt, Epsom salts and bathsalts are crystalline in form.

Further activity

Ask your local comprehensive school for a seed crystal of copper II sulphate, or potassium and aluminium sulphate (alum), with a small quantity of saturated solution of the same substance. Suspend the crystal in the solution from a cotton thread, and leave for some days. The children can then watch the crystal grow.

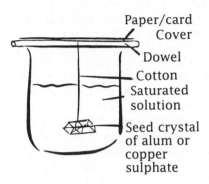

Paper/card Cover
Dowel
Cotton
Saturated solution
Seed crystal of alum or copper sulphate

Safety

Copper II sulphate is poisonous. The solution should be stored or disposed of safely. The enlarged seed crystal should be kept away from children. If they touch it they should wash their hands.

10. Separating colours: chromatography

Age range

Five to nine (first activity), and ten to eleven (second activity).

Group size

Individuals.

What you need

Filter paper or blotting paper, water soluble felt-tipped pens, Smarties, yoghurt pots, dropping pipette, water.

What to do

Ask the younger children to fold a piece of filter paper or blotting paper in two, and draw an outline of half a butterfly so that the fold is at the centre. They should cut out the shape of the butterfly and, using felt-tipped pens, colour a band about 1cm in from the fold on each side.

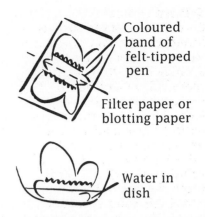

Coloured band of felt-tipped pen

Filter paper or blotting paper

Water in dish

The children should fold the butterfly and dip the fold in water, ensuring that the water does not touch the colour directly, and allow the water to soak upwards along the butterfly's wings, taking the colour with it, until it gives an effect they like.

For the second activity, the older children can put a Smartie in the centre of a round piece of blotting or filter paper and put this on top of a yoghurt pot. Ask them to predict what will happen if water is dropped on to the Smartie. Will the colour in the Smartie dissolve? What colours will be produced? The children should only add one drop of water at a time until the water line reaches the edge of the paper, or most of the colour has dissolved from the Smartie. Ask them to record their findings.

Dropping pipette

Smartie on filter/blotting paper

Yoghurt pot

Science content

Chromatography is a technique for separating small amounts of substances from a mixture by using the rate at which a particular substance moves along or through a medium. As colours are often made up of more than one dye, chromatography may be used to separate the individual colours. More than one colour should be obtained from most felt-tipped pens and Smarties, and older children should realise the potential of the technique for separating substances.

Further activity

The children can repeat these activities using food dyes and coloured inks. However, the colour from spirit-based pens will not separate unless a spirit-based solvent is used to dissolve the ink.

Relate chromatography to forensic science, and ask the children to investigate a forged cheque. Write a cheque on blotting paper, filter paper or ordinary paper, and ask them to investigate it by putting water on the ink and seeing which colours are produced. They can use this process to match the ink on the forged cheque with the ink in the pens of three suspects.

Safety

Care should be taken when using colours, even if they are water-soluble. The Smarties used for the experiment should not be eaten.

AT6/5c

11. Separating substances – a chemical solution!

Age range
Nine to eleven.
Group size
Small groups.
What you need
A mixture of sand and salt, two beakers, filter funnel, filter paper, spoon.
What to do
The children should put the sand and salt mixture into a beaker, add some water and stir well. They can then line a filter funnel with a piece of filter paper and put the funnel over another beaker. Next, they should pour the mixture of sand, salt and water through the funnel. What can they see left on the filter paper? What colour is the solution that has collected in the beaker?

Science content

The salt will dissolve in the water but the sand will not. The salt dissolved in the water passes through the filter paper and funnel as a clear solution, while the sand remains on the filter paper.

This is a method which can be used to separate two solids, one which is soluble (the salt) and one which is insoluble (the sand).

The soluble solid (salt) can be obtained by evaporation of the salt solution.

Further activity

Other mixtures could be separated in the same way; for example, sand and sugar. However, one solid must always be soluble and one insoluble.

AT6/5c

12. The water cycle

Age range
Ten to eleven.
Group size
Small groups.
What you need
Photocopiable pages 186 and 187.
What to do
Discuss with the children the water cycle, explaining such words as precipitation, evaporation and condensation. Most evaporation takes place in the sea, as it covers about 70 per cent of the Earth's surface, but water also evaporates from fields, plants, rivers and lakes, and even from our bodies. The children can next cut out the captions on photocopiable pages 186 and 187, which illustrate stages of the water cycle, and stick them on the relevant places on the diagram. It is important to emphasise that this process is a *cycle*.
Science content
Water is usually found as a

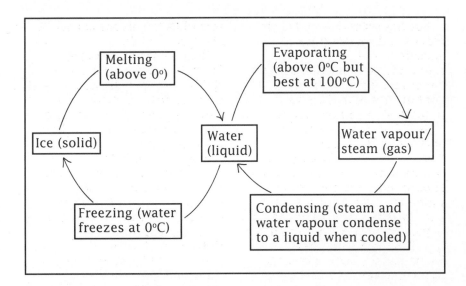

liquid but it can also be found as a solid (ice) and a gas (steam or water vapour). Nearly 98 per cent of the world's water is in liquid form, about 2 per cent is ice and very little (0.0005 per cent) is in the form of water vapour in the atmosphere.

Water vapour in the kitchen condenses on to the cold glass windows; this may be demonstrated in the classroom with a kettle. This process is similar to the way water vapour cools as it rises to form clouds, and falls as rain.
Safety
Hot steam, like the boiling kettle itself, can cause serious burns. Beware of children running into the flex of the kettle. Never use water near electrical items.
AT9/5c

CHAPTER 10

Earth and space

The mystery of space is a subject that has always fascinated people. How was the Earth formed? How do the Earth, Moon and Sun relate to each other? What do we know about other planets in our solar system? Why do we have four seasons? What causes an eclipse? You can use the interest generated by the topic of Earth and space to help the children to answer these complex and demanding questions.

Children should develop their knowledge and understanding of the relative positions and movement of the Sun, Earth, Moon, and the whole solar system within the universe. They should observe their local environment to detect seasonal changes, including day length, weather and changes in plants and animals. They should investigate the use of a sundial as a means of observing the passage of time.

ACTIVITIES

1. Keeping a diary

Age range
Five to six.
Group size
Individuals.
What you need
Paper and pencils.
What to do
Ask the children to keep a diary to show how the trees change during the year. They should look at the leaves, buds, flowers and seeds. They can also draw pictures to show any changes in the trees throughout the year.

This diary could be kept on a monthly basis starting in September and finishing in July of the school year.
AT16/1a
AT2/3c

2. Dressing for the occasion

Age range
Five to six.
Group size
Small groups.
What you need
Catalogues and magazines, large pieces of paper, pencils, scissors, adhesive.
What to do
Ask the children to look through the catalogues and magazines and cut out pictures of clothes they might wear in summer and in winter. What differences are there between the summer and winter

clothes? The children can then stick the pictures on to separate sheets of paper – one labelled 'summer clothes' and the other labelled 'winter clothes'.
Further activity
Children could group pictures of summer sports such as swimming, and winter sports, such as skiing, together with the different clothes they would wear for each sport.
Safety
Care must be taken when children are using scissors.
AT16/1a

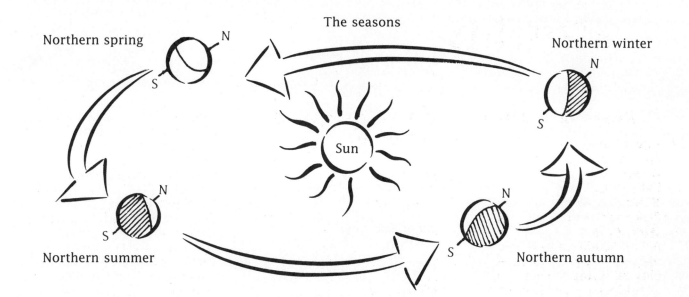

The seasons

Northern spring

Northern winter

Sun

Northern summer

Northern autumn

3. Four seasons

Age range
Five to six.

Group size
Small groups.

What you need
Four pictures showing different seasons, photocopiable page 188.

What to do
Look at the pictures on page 188 with the children. Which picture do they think shows winter, and which shows summer? How do they know? Ask them to name the other two seasons of the year and select the pictures which show the seasons. What differences can they see in the pictures?

Science content
The Earth takes one year to make a complete orbit of the Sun. The Earth's axis is tilted at 23.5° from the vertical, and so the North and South Poles lean towards the sun at different times of the year. When the North Pole is tilted towards the Sun, in June, the southern hemisphere experiences winter. Six months later, when the South Pole is nearest to the Sun, the northern hemisphere experiences winter.
AT16/1a

4. Under the Sun

Age range
Five to seven.
Group size
Small groups or pairs.
What you need
Thin tissue paper, small cut-out paper circles, Blu-Tack.
What to do
Cover a sun-facing window with a layer of thin tissue paper, and mark on the paper the positions of the Sun during the morning as seen through the paper. This can be done using small paper circles to represent the Sun and sticking them on with Blu-Tack.
Science content
The Sun appears to rise in the east and set in the west. This is due to the Earth's movement. One complete rotation of the Earth on its axis takes 24 hours. The Sun is low in the morning, high in the middle of the day and low again in the evening when it is in the west. It disappears from view at night.
Safety
Children must *not* look directly at the Sun. It can damage the eyes.
AT16/1c

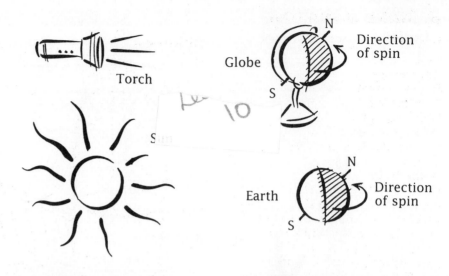

5. Night and day

Age range
Six to eight.
Group size
Pairs.
What you need
Torch, globe.
What to do
In a darkened room, ask the children to shine a torch on to a globe. Which side is dark and which is light? They can slowly rotate the globe in an anticlockwise direction to illustrate the progression of day and night as each country gradually moves into the light and then back into shadow.
Science content
The Earth makes one complete rotation on its axis every 24 hours. The Sun only shines on one side of the Earth at a time, so giving day and night.
Further activity
The children could make lists of nocturnal animals and animals which are active during the day.
AT16/2a

6. Make a shadow clock

Age range
Seven to nine.

Group size
Small groups.

What you need
Soft-drinks bottle, stick, sand, metre rule, chalk.

What to do
On a sunny day ask the children to place in the playground a bottle containing some sand, with a stick in it. They should draw round the base of the bottle with some chalk, and then draw round the shadow of the bottle and stick, and measure its length. The children can come back each hour and draw round the new shadow. What do they notice about the length of the shadow? When is it longest? In which direction does the shadow point?

Science content
The Sun has a diameter of 1,390,500km. It is a star at the centre of our solar system. The Earth is one of nine planets of the solar system which move in their own orbit around the Sun. The Earth takes 365.25 days to orbit the Sun. The position and size of a shadow will change throughout the day, and from day to day, because the Earth is orbiting the Sun and also spinning on its own axis once every 24 hours. In the summer, the shadow cast by the stick at a particular time of day is shorter than it is in winter.

Further activity
Make shadow clocks at various times during a year. Compare the lengths of the shadows in summer and winter.

Safety
Children must *not* look directly at the sun.

AT16/3a, 3b

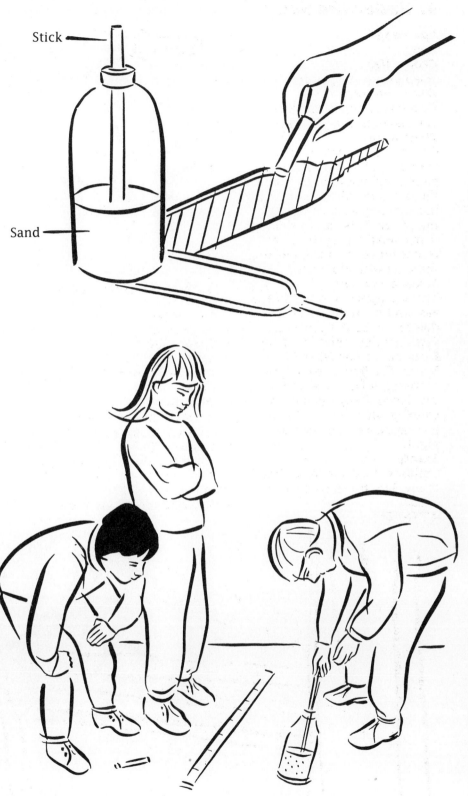

Stick

Sand

7. Me and my shadow

Age range
Five to seven.
Group size
Pairs.
What you need
Piece of chalk.
What to do
On a sunny day ask the children to stand in the playground and mark the position of their feet. They can then ask a partner to draw round the line of their shadows with a piece of chalk. They should do this regularly throughout the day, standing in the same position each time. When is their shadow longest? Which way does their shadow point?
Science content
Shadows are formed when some rays of light continue to travel in straight lines, while other rays are stopped by a person or object.
Further activity
Children can make shadow puppets or a puppet theatre.
AT16/3a, 3b

8. Lunar record

Age range
Nine to eleven.
Group size
Individuals.
What you need
Moon chart, photocopiable page 189.
What to do
Ask the children to look at the shape of the Moon over a period of a month and record the nightly shape of the Moon on the chart on photocopiable page 189. They should colour the bright part of the Moon yellow and write the date under each picture. If the Moon is hidden by clouds they should leave the circle blank.

How has the Moon's shape changed over the month?
Science content
The Moon shines because it reflects the light of the Sun. The Moon moves around the Earth approximately once every 28 days. As it does so, we see different amounts of the same surface lit by the Sun. That is why the Moon seems to change its shape during the month. The changing shapes of the Moon are called the 'phases of the Moon', and are shown in the chart on page 117.

Full Moons occur, on average, every 29.5 days. This timing is due to the combined effects of the Moon's orbit round the Earth and the Earth's orbit round the Sun.
Further activity
These records could be made as part of a project on space. Children could write about imaginary journeys to the Moon, and make a model of the surface of the Moon.
Safety
These observations should be made at home in the early evening during winter, with the help of the children's parents.
AT16/4a, 5b

New Moon: the Moon's Earth-facing side is dark	
Crescent: a thin crescent of light becomes visible.	
First quarter: the Moon now has half of its surface lit.	
Gibbous: the lit portion of the Moon is bigger than a semi-circle but smaller than a circle.	
Full: the whole of the Moon's Earth-facing side is lit.	
Gibbous: the Moon is on the wane, so that the sunlit portion becomes smaller.	
Last quarter: only half the Moon's surface is lit again.	
Crescent: there is now only a thin crescent of light visible.	

9. The Sun and the planets

Age range
Nine to eleven.

Group size
Ten children.

What you need
Cut-out pictures of the planets and the Sun, metre rules, reference books giving information on the Sun and the planets.

What to do
Ask the children the following questions:
• How many planets are there in the solar system?
• Which planet is nearest to the Sun?
• Which planet is furthest from the Sun?

Each of the children should pretend to be a planet. Using the scale of 1cm to represent 1 million kilometres, organise the planets in a line across the playground. Make sure that they stand in the right position in relation to the Sun.

How many planets are nearer to the Sun than the Earth?

Planet	Distance from the Sun (million km)
Mercury	58
Venus	108
Earth	150
Mars	228
Jupiter	778
Saturn	1427
Uranus	2870
Neptune	4497
Pluto	5941

Science content
There are nine planets in the solar system; Earth is one of them. The table above shows the approximate distances of the planets from the Sun. Distances are in millions of kilometres, but can be read as scale distances in centimetres.

Further activity
This activity could be combined with a visit to a planetarium.
AT16/4b

10. How big are the planets?

Age range
Nine to eleven.
Group size
Ten children.
What you need
Ruler, felt-tipped pens, scissors, large sheets of paper.
What to do
Using the information in the table opposite to help them with the size, ask the children to draw a picture of each planet and cut it out. They can then make a display of these pictures on the classroom wall. Ask them to find out as much information as they can about each of the planets. Which planets are smaller than Earth? Which planets are larger than Earth?
Science content
Compared to the Sun, all the planets are small. Pluto is the smallest planet and Jupiter is the largest planet.
Further activity
Combine this work with a visit to a planetarium.
AT16/4b

Planet	Approximate diameter in relation to the Earth (cm)
Mercury	0.4
Venus	1.0
Earth	1.0
Mars	0.5
Jupiter	11.0
Saturn	9.5
Uranus	3.7
Neptune	3.5
Pluto	0.5
Sun	109.0

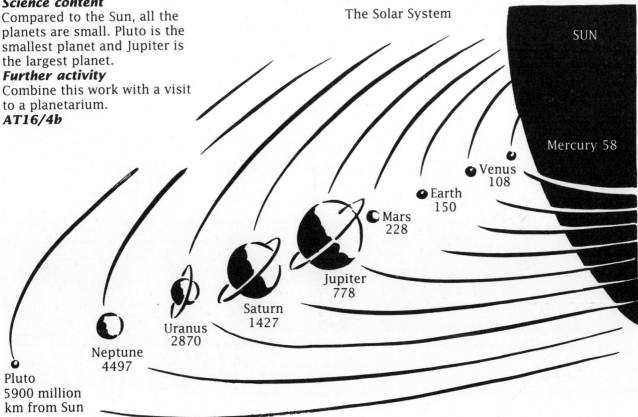

The Solar System

SUN

Mercury 58

Venus 108

Earth 150

Mars 228

Jupiter 778

Saturn 1427

Uranus 2870

Neptune 4497

Pluto 5900 million km from Sun

CHAPTER 11

Teaching guides

This chapter supports a cross-curricular approach to primary science, offering a range of topics which are rich in science. For each topic there is a planning web to show some possible lines of development, directly related to ideas from the National Curriculum science document.

Many of the science activities set out in this book can be built into the topics. Relevant activities are listed on the webs by number; for example, (3/10) denotes Chapter 3, Activity 10. Guidance has also been given on linking curriculum areas other than science to the topics. The chapter concludes with a grid linking the book's science activities to the science levels of attainment and attainment targets.

TOPICS

Machines

English:
• Creative writing – 'Invent a machine to ...'.
• Descriptions of machines in the home.
• Vocabulary – list the machines in the home.

Mathematics:
• Shape – look at cylinders.

• Turning.

Music and drama:
• Making machine sounds.
• Rhythms from machines.
• Turning.

RE:
• Working together.

History:
• Building the pyramids.
• History of machines.
• Industrial Revolution.

Geography:
• Farm machines.
• Visit to a farm or windmill.

Technology:
• Design and make a machine to....

Art:
• Making machines out of junk materials.
• Shapes – look at printing.
• Rotation.

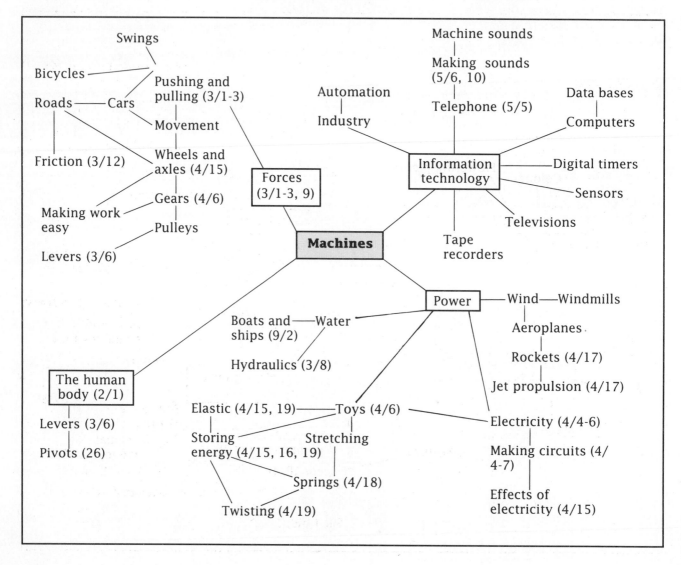

Plants

English:
• Books to read: *The Bongleweed*, Helen Cresswell (Puffin).
The Secret Garden, Frances Hodgson Burnett (Puffin).
• Creative writing: 'The Jungle'.

Mathematics:
• Growing and measuring plants.
• Symmetry.
• Shape.
• Plant database.

Music and drama:
• 'Harvest' songs.

RE:
• Food shortages, famine, Oxfam, Christian Aid.
• Harvest festival.

History:
• The feudal system.
• The history of plants such as wheat and potatoes.

Geography:
• Plants used for food, fuels, medicine.
• Coffee, tea, sugar, bananas.
• Exotic fruit and vegetables and their countries of origin.

Art:
• Seed collages, potato prints.
• Drawing plants.

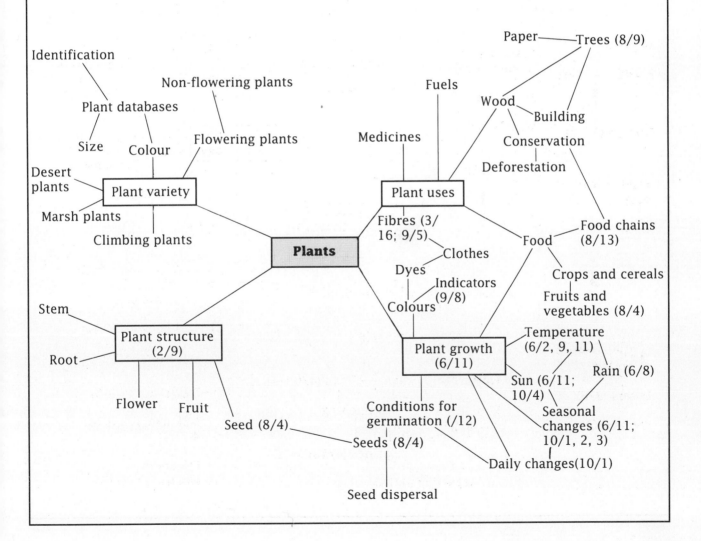

Pollution

English:
- Discussions on smoking and the conservation of resources.
- Advertising/posters – recycling, anti-smoking.

Mathematics:
- Litter surveys.
- Simple databases – 'Our facts'.
- Weighing rubbish in categories.

Music and drama:
- Fish in a polluted stream.
- Write a 'pollution' song.

RE:
- Discussions on food and famine, waste and poverty, use of energy resources, and air and water pollution.

History:
- Industrial Revolution.
- History of the Green Party.

Geography:
- Growing fruit and vegetables.
- Raw materials – mining and quarrying.
- Fuels – oil and coal.
- Farming and the loss of hedges.
- Country code.

Technology:
- Reducing pollution in power stations.
- Recycling.
- Lead-free petrol.
- 'Design a machine to ...'.

Art:
- Collage of an industrial skyline made from wrapping papers and scrap materials.
- Board game about water pollution.
- Posters.

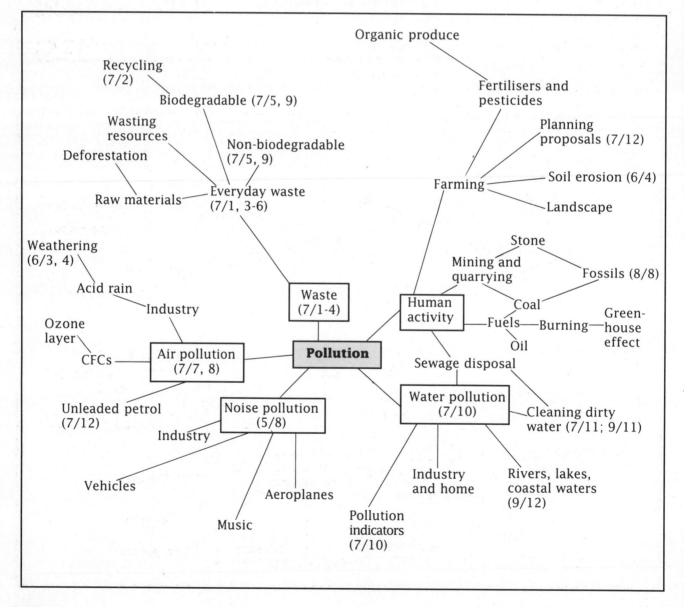

Communication

English:
- Writing letters.
- Discussions.
- Codes.
- Report writing.
- Advertising.
- Start a class newspaper.
- Computer software; for example, 'Front Page Extra', 'Folio' or 'Advanced Folio'.

Mathematics:
- Number.
- Counting games.
- Measuring.

Music and drama:
- Sounds, rhythms.
- Newspaper office.
- Mimes.

RE:
- Bible stories and parables.

History:
- The history of communication.
- The Post Office.
- The telephone.
- Flags, beacons.
- Transport system.
- Other communication systems.

Geography:
- International communication – air, roads, rail.
- Maps – signs, symbols and keys.

Technology:
- Telephone, television, facsimile, satellites.

Art:
- Lettering.
- Signs and symbols.
- Advertisements.
- Flags.

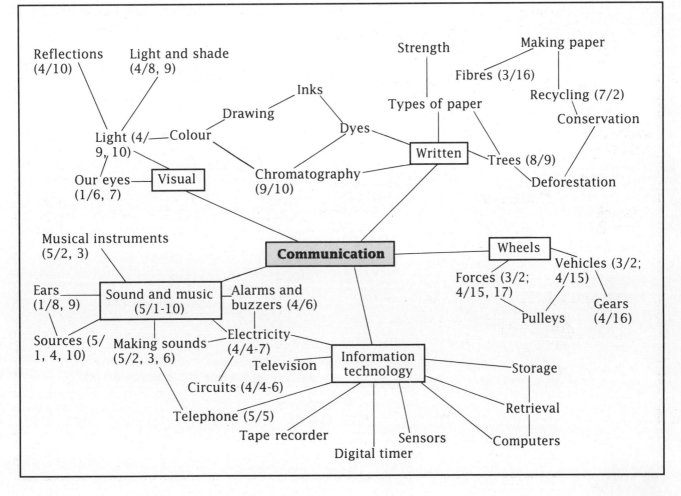

Transport

English:
• Creative writing – 'Fantastic journey' or 'A journey into space'.
• Letters.
• Travel brochures.
• Books to read: *Charlie and the Great Glass Elevator*, Roald Dahl (Puffin); *The Railway Children*, E. Nesbit (Puffin); *Around the World in 80 Days*, Jules Verne (various publishers).

Mathematics:
• Plans – look at children's journeys to school.
• Distances – how far and how long?
• Estimates and measurements.
• Traffic surveys.
• Timetables and fares for buses and trains.

Music and drama:
• Music and movement based on actions and sounds of forms of transport, for example, buses and trains etc.

RE:
• Journeys in the Bible.

History:
• History of flight.
• History of the bicycle.
• History of canals.
• History of the railways.
• Space.
• Famous bridges.

Geography:
• Travel round the world.
• Transport used in other countries, for example, canoes and high speed trains.
• River and canal transport.

Technology:
• Design and make powered vehicles.
• Make and test bridges.
• Design car park layout.

Art:
• Road safety and road signs.
• Large collage on forms of transport.
• Thomas the Tank Engine.

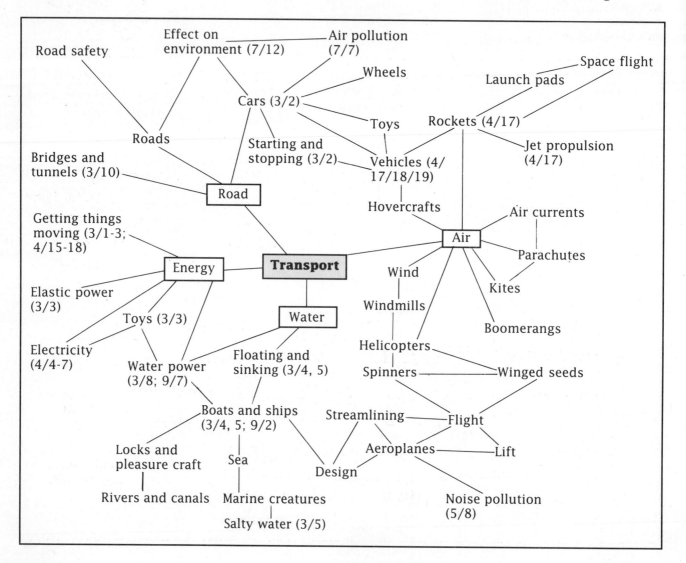

Light and colour

English:
• Writing advertisements and the use of colour.
• Logos.
• Photographs – talking about family photographs.
• 'Colour' words, such as warm and cold.
• 'Colour' walks.

Mathematics:
• Graphs of children's favourite colours.
• Symmetry and reflections.

Music and drama:
• Light and colour songs.
• Tone colour in music.

RE:
• Candles.
• Festivals of light, eg Diwali.

History:
• Inventions, eg electricity.
• History of photography.

Geography:
• Different times in different countries.
• Positions of stars.

Technology:
• Make a lighthouse.
• Lighting a puppet theatre.

Art:
• Painting, mixing colours.
• Making stained glass windows.
• Rainbows.
• Tie and dye.
• Colour in nature.
• Camouflage.
• Observing candles.
• Road signs.
• Warm/cold colours.

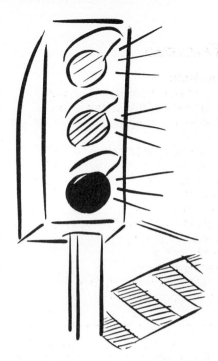

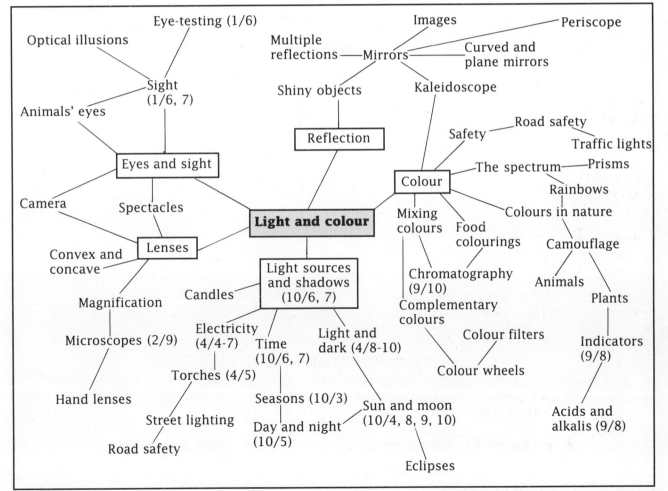

127

Houses and homes

English:
- Descriptions.
- Poetry.
- Creative writing – 'Homes of the future'.
- Estate agents' specifications.

Mathematics:
- Angles in buildings.
- Shapes in buildings.
- Plans and scale.
- Measurements of doors and gardens.
- Area.
- Graphs to show the types of houses the children live in.

Music and drama:
- Play – 'At home'.
- Rhymes.

RE:
- Homelessness.
- Noah's Ark.
- Churches.

History:
- Castles.
- Roman villas, arches.
- Famous houses.
- Tudor homes.

Geography:
- Homes in other countries.
- Scale and plans.

Technology:
- Design a house.

Art:
- Making models of houses.
- Collage of homes in other countries or animals' homes.
- Pictures about Hansel and Gretel.
- Pictures about the 'old woman who lived in a shoe'.

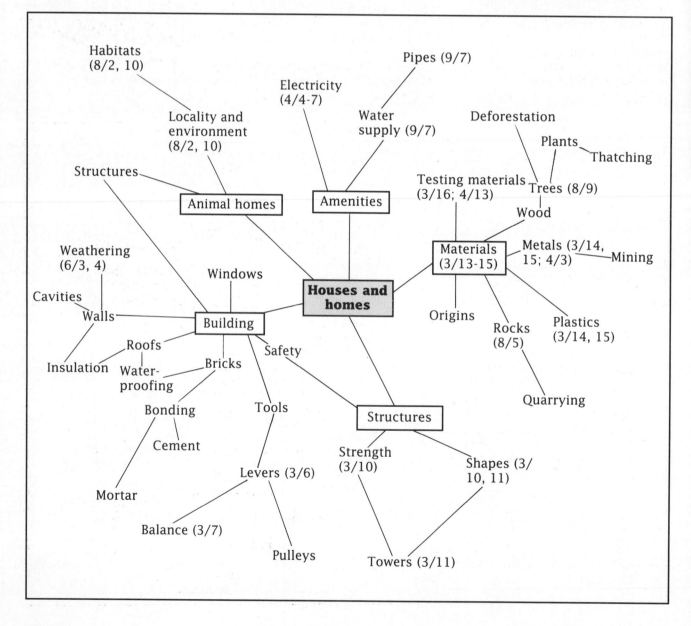

Shops and shopping

English:
- Advertising using posters.
- Vocabulary of goods sold in shops.
- Describing taste.
- Discussions about the layout of shelves and differences between small shops and large supermarkets.
- Interviews with people using shops.

Mathematics:
- Make a shop in the classroom – look at prices and buying and selling.
- Simple money transactions.
- Shapes of packets.
- Volume and capacity.
- Weighing.

Music and drama:
- Play about shopping.
- Counting rhymes.

RE:
- Feed the world.

History:
- Food in history.
- Markets and fairs.

Geography
- EC – trade and farming.
- Food from other lands.
- Everyday foods and goods – the corner shop, supermarket and department store.
- The high street.

Technology:
- Design packaging.

Art:
- Collage of food labels.
- Colours of fruits and vegetables.

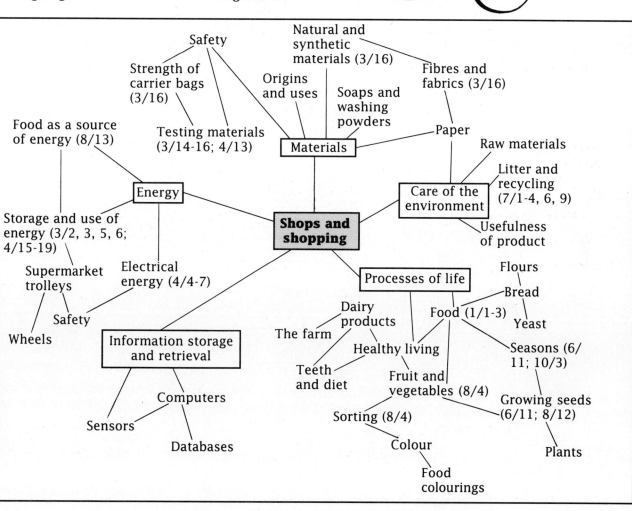

129

Christmas

English:
- What I do on Christmas Day.
- What Christmas means to me.
- Poems.
- Christmas crossword.
- Letters to Santa Claus.

Mathematics:
- Number patterns on a calendar.
- Grouping/classification of wrapping paper patterns.
- 'The Twelve Days of Christmas...'.

Music and drama:
- Nativity play.
- Party games.

RE:
- Christmas story.

History:
- Christmas long ago in Britain and other countries' Christmas customs.
- St Nicholas, Santa Claus.

Geography:
- North Pole.
- Jerusalem and Bethlehem.
- Where do reindeer live?
- Christmas round the world.

Technology:
- Making candles and baking.
- Model of Santa Claus moving up and down a chimney.
- Designing and testing wrapping paper.
- Designing and making the stable at Bethlehem.

Art:
- Painting the stable.
- Christmas collage.
- Decorations for tables and the whole room.
- Candles.
- Christmas trees, crackers.

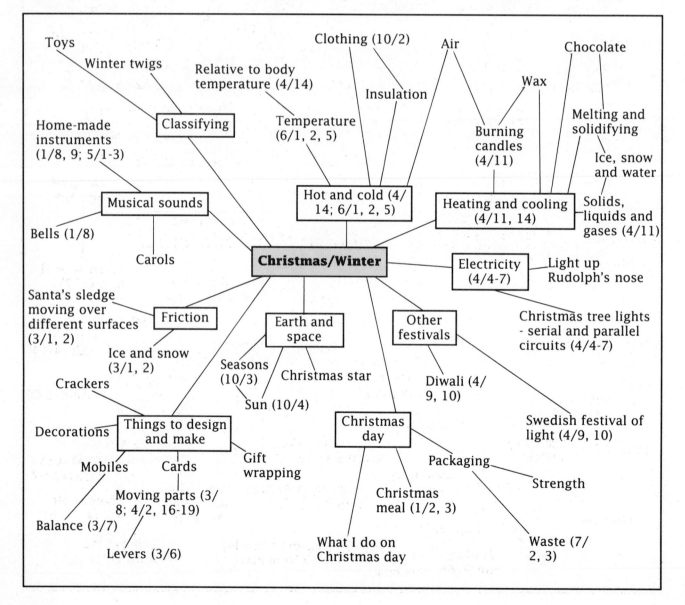

Ourselves

English:
- What I look like.
- My clothes and preferences.
- My family.

Mathematics:
- Graphs of favourite colours/food/television programmes.
- Height/hair/eye colour.
- Pulse rate/heartrate/breathing rate.

Music and drama:
- 'The knee bone connected to the thigh bone'.
- 'Simon says'.
- Clapping and tapping feet.

RE:
- Caring for others.
- Misuse of drugs.
- Friendship.

History:
- Evolution.
- Humans through the ages.

Geography:
- Children from other countries and climates.
- Homes in other countries.

Technology:
- Moving faces (levers).
- Medical advances.
- Stethoscope.

Art:
- Hand and finger painting.
- Fingerprints.
- Painting myself.

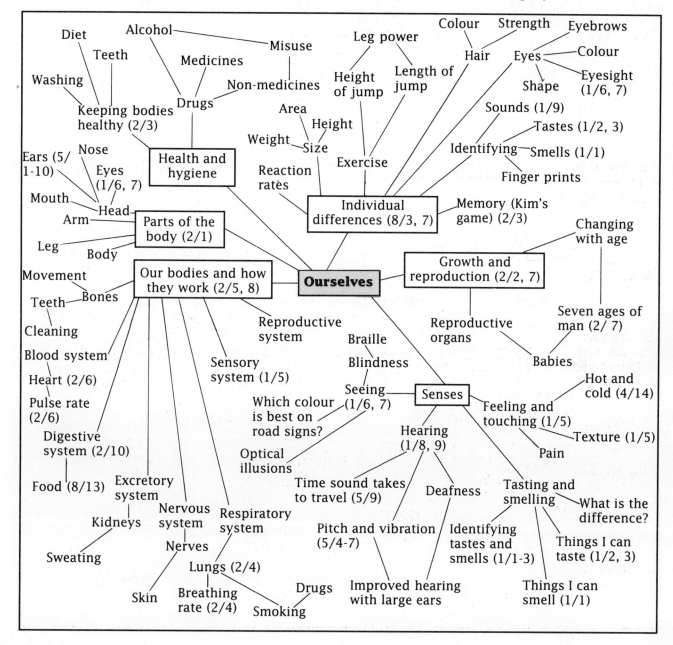

A T CHART

Use the following chart on this page and pages 135 and 136 to help link up the activities in this book with the science levels of attainment and attainment targets. The activities are identified by numbers; thus **8**/1 represents Chapter 8, Activity 1.

NB As AT1 is very general and is covered throughout this book, only a few examples of activities are listed on this chart.

AT Level	1 Exploration	2 Variety	3 Process	4 Genetics	5 Human influences
1 a	3/14 **10**/3	**8**/1 **8**/4	2/1	2/1	**7**/1-4 **7**/6
b	3/14 **10**/3				
c					
2 a	**8**/10	**8**/2 **8**/10 **8**/12	2/2 2/5	2/1	**7**/5 **7**/9
b	3/16 **8**/3	**8**/2 **8**/10	2/3 4/1		**7**/9
c	1/6 **8**/9		2/3		
d	3/4 **3**/13				
e	3/4-5				
f	**8**/3 **10**/2				
3 a	**7**/5 **8**/10	**8**/3-4 **9**/1	1/1-7 2/4-6	**8**/8	**7**/7-8 **7**/11
b	**7**/5	**8**/3-4	2/7		**7**/2 **7**/4
c	3/2 **8**/12	**10**/1			
d	3/16				
e	4/12				
f	6/8				
g					
h					
i	2/3 **8**/12				
4 a	3/12	**8**/3 **8**/7 **8**/11	2/6 2/8	**8**/9	**7**/2 **7**/6
b	**8**/12	**7**/5	2/8		
c	3/2 **8**/11-12	**8**/8			
d	3/2 **8**/11-12				
e	3/9				
f					
g	**7**/4				
h	6/8 -10				
i	2/6 **3**/7				
j					
5 a		**8**/10	2/9		**7**/7-8 **7**/10-11
b	**8**/12				**7**/9
c	3/9		2/10		**7**/12
d	**7**/8	**8**/13			
e					

AT Level		6 Materials	9 Earth and atmosphere	10 Forces	11 Electricity	12 Information technology
1	a	3/14	6/1	3/1-2		
	b		6/1			
	c					
2	a	3/14 3/16	6/1-2 6/6	3/1-3	4/2-3	
	b	3/13-14 3/16 9/3	6/2 6/11		4/4-5	8/4
	c	4/11				
	d					
	e					
	f					
3	a	3/13 3/16	6/3	3/8 4/18 9/2	4/5	
	b	3/13-14 4/11 9/3	8/6	3/4-5	4/4 4/6	8/4
	c		6/3-4			8/4
	d		6/4 8/5			
	e		6/5-6			
	f					
	g					
	h					
	i					
4	a	3/14-15 9/4-5	6/7-10 8/6	3/8 4/18	4/6	
	b	3/14-15 9/4-5	6/11	3/2		
	c	4/12				
	d	4/11 4/13		3/9		
	e	4/13				
	f					
	g					
	h					
	i					
	j					
5	a	8/6			4/7	
	b	9/8		3/10-11	4/15	
	c	9/10-11	9/12	3/2		
	d			3/12 4/18		
	e					

	AT	13 Energy	14 Sound	15 Light	16 Earth and space
1	a	4/1	1/8 5/1-2		10/1-3
	b	4/1			
	c				10/4
2	a	6/9	1/9	4/8	10/5
	b		5/2-3		
	c				
	d				
	e				
	f				
3	a	4/17-19	5/4-5	4/9	10/6-7
	b	4/11	5/6-9	4/9	10/6-7
	c	3/6-7 4/16-19			
	d				
	e				
	f				
	g				
	h				
	i				
4	a				10/8
	b				10/9-10
	c	4/17-19			
	d	4/11 4/14			
	e	4/11			
	f				
	g				
	h				
	i				
	j				
5	a		5/4	4/9-10	
	b		5/4		10/8
	c		5/8 5/10		
	d				
	e				

CHAPTER 12

Assessment, recording and evaluating

'Teaching, learning and assessment are interrelated. Assessment should form a natural part of teaching and learning activities.... Within the National Curriculum in England and Wales, teacher assessment should become an integral part of the teaching and learning process.' (SEAC 1990). This chapter is aimed at helping you, the teacher, to fully appreciate the nature and purposes of assessment. What are the methods of assessment available to you in the classroom? What is the purpose of record keeping and evaluation?

ASSESSMENT

Assessment involves you in gathering information from a child's interaction with an aspect of scientific activity. This information is compared with a predetermined standard or level. The standard or level referred to may be loosely defined through experience, as when work is marked with statements such as 'good' or 'poor', or marks such as 8/10, or a frown or a nod of appreciation. Alternatively, the standard may be more tightly defined, perhaps using previously determined criteria agreed among colleagues or suggested by another source.

Assessment: some definitions

Michael Bassey in his *Assessment guide for Primary Schools* (1990) uses and defines the following useful terminology associated with assessment:

• *Formative assessment* helps inform the teaching process. It is assessment during teaching which enables you to determine what future actions are appropriate in your classroom, for example the work which particular pupils should carry out on the evidence of their present performance.

• *Summative assessment* concludes a period of teaching and learning in order to determine what has been learned.

• *Informal assessment* arises during the normal day-by-day routine in your classroom; it is based on your informal observation of pupils at work in the classroom, when they answer questions and discuss investigations, on the items they make and when reading their work. From this wealth of information, judgements and achievements of pupils against the Statements of Attainment may be made. This type of assessment involves the professional judgement of you the teacher.

• *Formal assessment* arises from *specially* prepared tasks which provide tangible evidence on which judgements can be made.

With regard to the National Curriculum it can be seen that while Teacher Assessments (TAs) are informal, producing both formative and summative outcomes, Standard Assessment Tasks (SATs) are formal, giving rise to summative outcomes.

Why assess?

Despite its obvious importance, the monitoring, recording and evaluating of children's work in primary science has suffered serious neglect. There are two main reasons for assessing children's work:
• You need to know the levels of ability and progress of the children arriving in your classroom, so that the programmes of study and schemes of work may be translated into activities in the classroom for the whole class, groups of pupils or individuals.
• You need to provide records of progress. These will be based on data collected through assessment.
You interact with your pupils throughout the day as questions are asked, help is required, conversations are overheard, work is brought for checking and discussions about a particular investigation take place. Without perhaps being aware of it, you carry around in your head a vast amount of information on the abilities and progress of the children and the problems and difficulties encountered in their learning.

Sometimes this information is transferred to a written record of the children's individual progress. The worth of these records will depend on the methods you have used to obtain the information and the details you give about the information so obtained. As a teacher and an assessor, you need to be clear about what you are assessing and the criteria upon which that assessment is carried out, so that the information you obtain can help you to further the children's learning more effectively.

The majority of methods of assessment, however they are carried out, are not 100 per cent accurate. Assessment involves making a judgement based on evidence which is then recorded formally on paper or informally in your mind. This judgement is based on your experience, honesty and professionalism as a teacher, and it is important to ensure that your bias or prejudice is squeezed out at the moment when judgement is made.

It is our duty to make judgements as accurately as possible, on the evidence before us. Teachers, of course, make many judgements every day, but when assessments need to be placed on a more formal footing in the form of written records, some teachers feel less secure about committing their judgements to paper. They argue that their judgements might not be totally accurate, or that the pupil might improve later on, or even that it is not their role to make such assessments and to keep records. Yet surely it is our role. It may help to remember that records should be flexible enough to be updated and modified, so that no judgement about a child needs to be accepted as accurate for all time. By adopting a rigorous, honest approach to record keeping we can, and must, live securely with our roles as assessors.

Assessing in the classroom

As science involves the methods of finding out information as well as the knowledge and understanding these methods disclose, it would be logical to assess children's exploration (process) skills as well as their knowledge and understanding. In addition, it is necessary to assess the attitudes which help children to learn about science.

A variety of methods are available for the teacher; some of these are analysed below for their merits and disadvantages.

Questioning
By asking both open and closed questions, you can gain a wealth of information about your pupils' knowledge and understanding. The flexible nature of one-to-one or group questioning allows for further questions to be asked to check what was really meant by an initial vague response. This is not possible with set tests, and is difficult to achieve with written work. Clearly, open questions allow for a greater range of responses than closed questions, and this may have more assessment potential. However, closed questions will generally be easier to assess.

Discussion
In discussion, children have a chance to tell us what they have been doing and why they have been doing it. Ideas, opinions and concerns are raised, and questions are asked and answered. It may initially be difficult to assess individual contributions to discussions, although this will improve with practice as you gain confidence.

Listening
Listening can involve the teacher in a passive role, avoiding the temptation to interrupt with questions or statements. However, in making judgements on the child's learning you will be actively listening, and you will have a chance to refer to successes or misconceptions later, as a prelude to matching work more appropriately. In addition, you may have gained assessment information which can be added to the records when time allows.

Structured observation
This allows the teacher to observe closely the ways in which the children are working. It overlaps with 'Listening' above, but the emphasis is on watching children's behaviour and looking for signs of interaction, co-operation, perseverance, responsibility, sharing, etc. It also gives a chance to see how children handle apparatus and how they tackle investigations and problems. Such observations enable assessment of processes, but although they take place during normal classroom activities, they are nevertheless time-consuming.

The number of children you can observe at any one time is normally limited to a maximum of three or four. You will, however, gain much useful information on how children deal with techniques such as taking temperatures, or how they handle experimental variables. Time should be set aside for carrying out structured observations, using a check-sheet which has room to note unexpected outcomes.

Interruptions are bound to occur, but you can return to the observation once the necessary assistance has been given.

This also raises a most important point, and one which concerns many teachers. Opportunities for and evidence of positive achievement occur in our classrooms on numerous occasions every day. We must ensure that we collect some of these examples but it is impossible to collect them all. Some remain to be found when we look at pupils' work, but some, and probably many, will apparently be lost. But do not worry, they will surface again at a later date in another context. The frenetic teacher galloping around the classroom sweeping in assessment data will be a liability, not an asset, to teaching and learning in the classroom. What is required is a well-planned strategy which ensures that assessments are made regularly during, and maybe after work on a topic or theme. The teacher who simply makes a few arbitrary assessments prior to the onset of Standard Assessment Tasks cannot be defended; a better way of working needs to be found.

Analysing children's work

Children's work can be looked at as the end product of an activity, and assessing it is part of your role; however, modifications may need to be made to the way in which it is tackled. While comments such as 'good' and 'neat work', and marks such as 8/10 may still have a place, indication of positive achievement may need to be made explicitly, particularly if the pupils are to be involved more in the teaching/learning/assessment process, which they surely should and must. Analysis of children's work should give you valuable information about their exploration skills, and particularly their recording skills, but it will also be a source of evidence for aspects of their conceptual understanding. Much of your children's work will be in written form, dealing for example with what they did and found out, and with diagrams and graphs, but alternative forms of recording should also be encouraged, such as tape recordings, drama, or the making of experimental set-ups or models.

139

Written tests

So far, we have indicated how both formative and summative assessments may be made in the classroom. Test-type information is formal in nature, but it may give both formative and summative outcomes. On the whole, however, written tests have tended to be used for summative purposes, except where remediation programmes have been set up following such tests. Tests are, of course, relatively easy to set, particularly if the response to each question is limited to one or two words. In this case, factual information is most easily tested, and it can be quickly assessed. However, there are problems with written tests; some children have difficulty with reading, the language used in tests is always problematic, and written tests may not be appropriate to assess exploration skills, particularly if they are out of context. For example, it is better to offer pupils the opportunity to measure the volume of water in an actual measuring cylinder than to give them a diagram of such an experimental set-up.

Formal testing in British primary schools has decreased over the last decade or two and has rarely, if ever, been a part of primary science assessment, but the introduction of Standard Assessment Tasks as an important part of the National Curriculum may lead some teachers to consider tests as a method of assessing and monitoring progress. If tests do become important in science assessment, they will need to be written carefully. The limitations of this method, like any other, must be clearly understood.

Criteria and assessment

Teacher assessment in the context of the National Curriculum is not based on a traditional testing model but is criterion-referenced. Thus, according to SEAC (1990), 'a child is assessed in relation to a criterion given by a Statement of Attainment and not in relation to other children'. This is welcome news, but it requires far closer scrutiny. It presupposes that the Statements of Attainment *are* criterion-referenced statements, which many of them are not, at least because they are written without explanation! Taking, for example, Attainment Target 14: Sound and Music, let us see how easy the statements are to interpret.

The statements of attainment as they stand are clearly not always criterion-referenced. For the time being, at least, it is unlikely that significant changes will be made, or that each statement will be clarified by the assessment body (SEAC). It is therefore up to teams of teachers to clarify for themselves what these statements really mean and what is required for assessment. Through sharing and critical analysis of assessed pieces of work, a consensus will arise. Whether the idea gained is the same as in the school in the next village or in another part of the country remains to be seen. There is still an open question as to how 'national' the assessment of the National Curriculum will be!

Level	Statement of attainment	Comment
1	Know that sounds can be made in a variety of ways	How many ways? How much variety?
2a	Know that sounds are heard when the sound reaches the ear	This is clear enough to interpret
3b	Be able to give a simple explanation of the way in which sound is generated...	What does 'simple' mean?
4	Know that it takes time for sound to travel	This is clear enough to interpret; speed not required
5c	Understand the importance of noise control in the environment	Not sufficient to repeat statement, but what 'understanding' is required?

RECORDING

'Recording information about pupils' achievement and progress, although important, is only part of the assessment process and not an end in itself.' (SEAC, 1990).

You will therefore need to keep records which include details of the scheme of work and programmes of study covered, but which serve the following purposes:
• to record the attainment targets and statements of attainment which each child has covered;
• to record the level of attainment which each child has reached for each attainment target;
• to provide the evidence to support the levels of attainment recorded;
• to give parents, other teachers and schools access to information about academic and other achievements in school.

An effective record system

If a record system is to be effective and useful it must successfully fulfil the following criteria.
• It should be simple and manageable to complete, so that it does not detract from the teaching and learning taking place in the classroom.
• It should include all the information required to further learning opportunities in the classroom and to fulfil conditions laid down by school, local or national bodies.
• It should enable pupils to understand the progress they are making.
• It should enable others who read it to gain the information they require.
• It should be filled in on the basis of evidence and not conjecture.

A whole-school recording policy and system is clearly required. Many primary and secondary schools within 'families' or 'pyramids' are working on joint record systems which will make the transfer of children's achievements more efficient and worthwhile. We suggest the following possibilities:
• a whole-class teaching plan and record;
• class record sheets;
• individual children's records.

A whole-class teaching plan and record will inform the teacher in detail of the work to be undertaken on a weekly basis, and may be used for monitoring purposes at a later date. For example, teachers would be able to tell whether they have covered a certain aspect of the programme of study, or given pupils an activity which allows them to indicate whether they have achieved a particular statement and level of attainment.

An example modified from the SEAC Assessment Pack C is shown below.

Class 6M	Week ending 25/6		
Curriculum area	**Details of work to be undertaken**	**Statement of attainment**	**Programmes of study**
Science	Visit to a windmill		
	Gears as power sources	13 SC 3a,3c,4c	Page 70
	Making a windmill	10 SC 3a,4a	

Science	Level 2														
Names	PC1						PC2 (Profile component)								
	AT1						AT2		AT3			AT4...	AT16		
	a	b	c	d	e	f	a	b	a	b	c	a	a	b	c
Muntu A.	✓	✓	✓				✓	✓	✓	✓	✓	✓	✓		
Jason B.	✓	✓	✓				✓		✓				✓		
Anna B.	✓	✓	✓				✓	✓		✓	✓	✓	✓	✓	
John C.	✓	✓	✓	✓	✓	✓	✓	✓	✓	✓	✓	✓	✓		
Praka D.	✓	✓					✓		✓			✓	✓		

Class record sheets may contain all the information for one level (see above).

While giving summative information on the progress of individual children and of the whole class, this type of record lacks the space to include such details as the context of the activity, when positive achievement was obtained, what evidence there was for the achievement or possible ways to help children in the future. By 'crossing' the tick, a distinction could be made between experiencing a particular statement of attainment and gaining positive achievement or mastery.

Individual children's records may be of two types.

• The first type is an individual record sheet, covering all the statements of attainment for an individual pupil. Each box could be cross-hatched when experienced, and coloured in when achieved. Different colours could be used for each year.

This may be used as a summative record for reporting, and also as a formative record for advising further action for the child, though there is no room to record the information on which it is based, such as the context, or the evidence of attainment. This system has the advantage of using only one piece of paper, but it lacks detailed information.

• The second type of individual record sheet (shown opposite) is more detailed, addressing each attainment target separately.

In this record sheet there is more space for information about the nature of the achievement, but this does of

---

Individual science record sheet
Key stages 1 and 2   Name.................................

Reception......................... ⎫
Yr. 1................................. |
Yr. 2................................. |
Yr. 3................................. ⎬ Teachers' initials and indicator
Yr. 4................................. | colour for each year
Yr. 5................................. |
Yr. 6................................. ⎭

AT / Level	1	2	3	4	5	6	9	10	11	12	13	14	15	16
Level 1	a	a	a	a	a	a	a	a	a	a	a	a	a	a
	b						b				b		b	b
														c
Level 2	a	a	a	a	a	a	a	a	a	a	a	a	a	a
	b	b	b		b	b	b		b	b	b	b	b	b
	c		c			c	c							c
	d						d							d
	e													
	f													
Level 3	a	a	a	a	a	a	a	a	a	a	a	a	a	a
	b	b	b		b	b	b	b	b	b	b	b	b	b
	c	c					c				c	c		
	d						d							
	e						e							
	f													
	g													
	h													
	i													
Level 4	a	a	a	a	a	a	a	a	a	a	a	a	a	a
	b	b	b			b	b	b	b		b		b	b
	c	c	c			c	c		c		c			c
	d		d				d		d		d			
	e						e				e			
	f													
	g													
	h													
	i													
	j													
Level 5	a	a	a	a	a	a	a	a	a	a	a	a	a	a
	b	b	b		b	b	b	b	b	b	b	b		b
	c	c	c			c	c	c	c	c		c		
	d	d	d						d					
			e											

142

Science AT 1 Exploration of science		Name		
SoA	Notes: Where/when covered; evidence; method of assessment; other.	Date 1	Date 2	Achieved
1sc1a	Topic on clothes, sorted fabrics into groups	14/9		14/9
1b	Able to tell friends what she had done	14/9		14/9
2sc1a	Asked why some fabrics are rough	14/9		
1b	Described fabrics as rough/smooth	14/9		

course take more paper. This sheet would need to be combined with the class record sheet, in order to analyse and summarise assessment information on the whole class.

Evaluation

Having established what is to be taught and assessed in science, worked out the criteria for assessment, assessed the pupils' work and recorded their achievements, you can evaluate the children's progress, the science scheme of work and your own teaching. You might like to ask yourself the following questions at the end of the topic, term or year.
• Did I complete the work which I set out to teach?
• Did some pupils accomplish less than I expected?
• Was the work well matched to the pupils' abilities?

• Were the pupils given the opportunity to achieve the statements of attainment? (Look to see if they were covered on your 'whole-class teaching plans'.)
• Was there an area that the children did not understand, even thought it was covered? How might I introduce it more effectively next time?
• Were my records useful to further children's learning and to inform parents, children and teachers about progress?

There is little need to be anxious about assessment and record keeping, as the vast majority of teachers are already assessing their children throughout the day. What you now need to do is to put it on a more formal basis, as already discussed. With a small amount of practice, teacher assessment will become a normal part of the teaching and learning activities in your classroom.

Some National Curriculum terms

• Attainment targets set out the knowledge, skills and understanding that pupils are expected to master as they progress through school.
• The key stages are the periods in each pupil's education to which the National Curriculum applies. There are four stages: Key Stage 1 – from the beginning of compulsory education to age seven; Key Stage 2 – ages seven to eleven; Key Stage 3 – ages eleven to fourteen; Key Stage 4 – age fourteen to the end of compulsory education.
• Levels of attainment – there are ten of these within each attainment target
• The programmes of study are the matters, skills and processes which must be taught to pupils during each Key Stage in order for them to meet the objectives set out in the attainment targets.
• Standard assessment tasks (SATs) are externally-prescribed assessments.
• Statements of attainment are more precise objectives than the attainment targets, and are related to each of ten levels of attainment.
• Teacher assessments (TAs) are internal assessments which are carried out by the teacher.
• The profile components are groups of attainment targets. In science there are two profile components:
1. Exploration of science, communication and the application of knowledge and understanding.
2. Knowledge and understanding of science, communication, and the applications and implications of science.

CHAPTER 13

Resources and books

No teacher can successfully teach primary science without an adequate supply of resources. These resources may be classroom- or school-based, but a combination of the two is best. Some specialist equipment will be needed, such as a newton meter or a light box, but so will some everyday resources, such as yoghurt pots and other junk items. It is important to realise that although some very successful primary science can be carried out with collections of junk materials there is a definite need for specialist equipment and other appropriate resources in order to teach many parts of the National Curriculum for science.

RESOURCES

Specialist equipment

Resources can be gradually built up over a period of time, although some purchases such as batteries, bulbs and spirit-filled thermometers will need to be made on a regular basis.

The purchasing of resources needs to be controlled and carefully planned in order to identify school and classroom needs within the budget available. Look at your existing resources first – there may well be some very useful equipment lying unused in a store cupboard, or you may have some 'useless' equipment which you could trade with a secondary school.

Compare the prices of equipment if you have the time, and try buying in bulk, perhaps with other schools in the local area. Think about your science teaching programme over the next two years and match your purchases against future needs throughout the school. You must not expect to buy all you need immediately – it will take time.

Think about storage of the equipment, as the resources should be readily available in stacking units, on shelves, in trays, or if you are lucky enough, in your science resources area or science room. Who is going to collect the equipment from the science resource area? Will it be the children or the staff? Whatever you decide, all storage units should be clearly labelled and easily recognisable to ensure ready access to equipment.

Everyday materials

There follows a list of everyday items which you may find useful in science activities. Also listed are a number of possible uses for each item.
• Balloons can become a monorail, vehicle or hovercraft.
• Bottle tops can be used to make gears.
• Carpet scraps and carpet tiles can be used as surfaces to test for friction.
• Cereal packets can be used for the cardboard (cut, turn inside out and reassemble, thus allowing painting).
• Coat-hangers can be made into mobiles or electric buzzer games.
• Cotton reels can be used to make wheels or a crawler.
• Drinks cans can be used to make a can roller, or as a container for insulation experiments.
• Egg boxes can become seats on vehicles or fairground rides.
• Jam-jars can be used as an alternative to beakers or for observing soil components.
• Lids can be used as wheels.
• Magazines, catalogues and calendars can be used in classification exercises or cut up for pictures.
• Margarine tubs, yoghurt pots and plastic cups can be used for growing seeds, instead of beakers, or as water timers.
• Newspapers and waste paper can be used as insulation or for building towers and bridges.
• Paper-fasteners will attach the moving parts of models.
• Plastic drinks bottles (1 litre) can be used as a rain gauge or a stand for a shadow clock, or can be cut up to make blades for windmills and water mills.
• Plastic spoons can be used to stir things, measure powders, or collect minibeasts.
• Polystyrene can be used to make windmills, as surfaces for friction, as parts of a wind vane and as sound insulation.
• Shoe boxes can be turned into musical instruments, feely boxes or a puppet theatre.
• Straws can be used in construction or as pipettes.
• 35mm film holders can be used as battery boxes.
• Insides from toilet rolls or kitchen rolls can be used as chimneys or vehicles, cylinders, tubes, towers or battery holders.
• Washing-up liquid bottles can be used to make windmills, crawlers or rollers.

- Waste fabrics can be used to look at fibres, as insulation, and to test different materials.
- Wine corks can be used to test for sinking and floatin,g or as holders for pins when testing skin sensitivity.
- Wood off-cuts can be used when testing for floating and sinking, or for sorting materials or buggy-making.

Useful addresses

Educational associations
- The Consortium of Local Education Authorities for the Provision of Science Services (CLEAPSS) produces primary guides which are free to schools in England and Wales whose LEAs belong to the consortium. The guides cover a useful range of subjects, including the following:
Small Mammals – Teaching Notes;
Safe Use of Household and Other Chemicals;
Advice on Microscopes and Magnifiers for Junior and First Schools;
Plants for Primary Science;
Storage for Primary Science;
Housing Animals;
Electrical Safety;
Tools and Techniques;
Batteries etc – Which to Buy?
Models of Humans, Animals and Plants;
Simple Electric Circuits with Bulbs and Batteries;

Elementary Photography;
Measuring Equipment;
Thermometers for Primary Schools;
Magnets;
Heating in Primary Science;
Environmental Equipment;
Construction Kits;
Glues and Adhesives.
    CLEAPSS can be contacted at the following address: CLEAPSS School Science Service, Brunel University, Uxbridge, UB8 3PH.
    In Scotland the equivalent service is the Scottish Schools Equipment Research Centre, 24 Bernard Terrace, Edinburgh EH8 9NX.
- The Association for Science Education is an organisation run by teachers for teachers, and it provides a national and regional forum for primary teachers. You can join either as an individual teacher or as a school subscriber, and ASE subscriptions are deductable for tax purposes.
    ASE members receive *Primary Science Review* (a bi-monthly magazine) and *ASE Primary Science* (a broadsheet containing ideas for classroom activities). They also get a ten per cent discount on many educational books, and the association runs its own specialist bookshop. Its own publications include *The National Curriculum: Making it work for the Primary School; Be*

Safe! Safety document for primary schools; and *Choosing published primary science materials for use in the classroom.*
    The association also organises many regional and national activities, including lectures, courses, workshops and exhibitions.
    ASE can be contacted at the following address: The Association for Science Education, College Lane, Hatfield, Herts AL10 9AA.

Museums
- Ironbridge Gorge Museum, The Wharfage, Ironbridge, Telford, Shropshire.
- Natural History Museum, Cromwell Road, London SW7 5BD.
- Science Museum, Exhibition Road, South Kensington, London.
- The Museum of Science and Industry, Newhall Street, Birmingham B3 1RZ.
- The Museum of Science and Industry, Liverpool Road, Castlefield, Manchester M3 4JP.

The environment
- Aluminium Can Recycling Association, Suite 308, 1 Mex House, 52 Blucher Street, Birmingham B1 1QU.
- Business and the Environment Unit, Department of Trade and Industry, Room

1016, Ashdown House, 123 Victoria Street, London SW1E 6RB.
• Council for Environmental Education, University of Reading, 24 London Road, Reading RG1 5AQ.
• Friends of the Earth, 26-28 Underwood Street, London N1 7JQ.
• National Agricultural Centre, Stoneleigh, Kenilworth, Warwickshire CV8 2LZ.
• National Association for Environmental Education, West Midlands College of Higher Education, Gorway, Walsall, West Midlands WS1 3BD.
• Nature Conservancy Council, Publicity Services Branch, Northminster House, Peterborough PE1 1UA.
• Royal Society for Nature Conservation, The Green, Nettleham, Lincoln LN2 2NR.
• Royal Society for the Protection of Birds (RSPB), The Lodge, Sandy, Bedfordshire SG19 2DL.
• The Department of the Environment, Wildlife Division, Tollgate, Houlton Street, Bristol BS2 9DJ.
• The National Society for Clean Air and Environmental Education, 136 North Street, Brighton BN1 1RG.
• The Tidy Britain Group, The Pier, Wigan WN3 4EX.
• The Wildfowl Trust, Slimbridge, Gloucestershire GL2 7BT.

• Urban Wildlife Group, 131-133 Sherlock Street, Birmingham B5 6NB.
• Waste Watch, National Council for Voluntary Organisations, 26 Bedford Square, London WC1B 3HU.
• World Wide Fund for Nature, Panda House, Weyside Park, Catteshall Lane, Godalming, Surrey GU7 1XR.

Earth and space
• Jodrell Bank Science Centre and Tree Park, Macclesfield, Cheshire SK11 9DL.
• Liverpool Museum and Planetarium, William Brown Street, Liverpool L3 8EN.
• Royal Astronomical Society, Burlington House, Piccadilly, London W1V 0NL.
• The Association for Astronomy Education, c/o The London Planetarium, Marylebone Road, London NW1 5LR.
• The London Planetarium, Marylebone Road, London, NW1 5LR.

Other useful addresses
• British Waterways, Canal Office, Delamere Terrace, London W2 6ND.
• Pictorial Charts Educational Trust, 27 Kirchen Road, London W13 0UD.
• Shell Education Service, Shell UK Ltd, Shell-Mex House, Strand, London WC2R 0DX.
• The Design Council, 28

Haymarket, London SW1X 4SU.
• Primary Schools and Industry Centre, The Polytechnic of North London, School of Teaching Studies, Prince of Wales Road, London NW5 3LB.

Some scientific and technological suppliers
• E. J. Arnold, Parkside Lane, Dewsbury Road, Leeds LS11 5TD.
• Commotion, 241 Green Street, Enfield EN3 7TD.
• Griffin & George, Bishop Meadow Road, Loughborough, Leicestershire LE11 0RG.
• Philip Harris Ltd, Lynn Lane, Shenstone, Staffordshire WS14 0EE.
• Heron Educational, Unit 12, Kenilworth Works, Denby Street, Sheffield S2 4QN.
• Hestair Hope Limited, St Philip's Drive, Royton, Oldham OL2 6AG.
• Nottingham Educational Supplies (NES), Ludlow Hill Road, West Bridgford, Nottingham NG2 6HD.
• Osmiroid International Ltd, Fareham Road, Gosport, Hampshire PO13 0AL.
• Sheffield Purchasing Organisation (SPO), Sheffield City Council, Staniforth Road, Sheffield S9 3GZ.
• Technology Teaching Systems Ltd (TTS), Penmore House, Hasland Road, Chesterfield S41 0SJ.

BOOKS

Useful books for the classroom

• The 'Conserving Our World' series, published by Wayland.
• The 'Look at' series, published by Franklin Watts.
• The 'Life Cycles' series, published by Wayland.
• The 'Secrets of Science' series by Robin Kerrod, published by Cherrytree Press.
• The 'Eyewitness' series, published by Dorling Kindersley.
• The 'Human Body' series, published by Franklin Watts.

Useful reference books

• British Association for the Advancement of Science *Ideas for Egg Races and other Problem Solving Activities.*
• Harlen, W. (Ed) (1985) *Primary Science : Taking the Plunge* (Heinemann Educational).
• Harlen, W. and Jelly, S. (1990) *Developing Science in the Primary Classroom* (Oliver and Boyd).
• Johnsey, R. (1987) *Problem Solving in School Science* (Macdonald Educational).
• Ross, A. *et al* (1990) *The Primary Enterprise Pack* (PNL Press).

REFERENCES

• Bassey, M. (1990) *Trent Assessment Guide for Primary Schools, National Curriculum Key Stage One* (Local Education Authority Publications).

• Department of Education and Science (1983) *Assessment of Performance Unit Science Report for Teachers 1: 'Science at Age 11'* (HMSO).

• Department of Education and Science and the Welsh Office (1989) *Science in the National Curriculum* (HMSO).

• School Examinations and Assessment Council (1990) *A Guide to Teacher Assessment, Pack C; A Source Book of Teacher Assessment* (Heinemann Educational).

The pages in this section can be photocopied and adapted to suit your own needs and those of your class; they do not need to be declared in respect of any photocopying licence. Each photocopiable page relates to a specific activity in the main body of the book and the appropriate activity and page references are given above each photocopiable sheet.

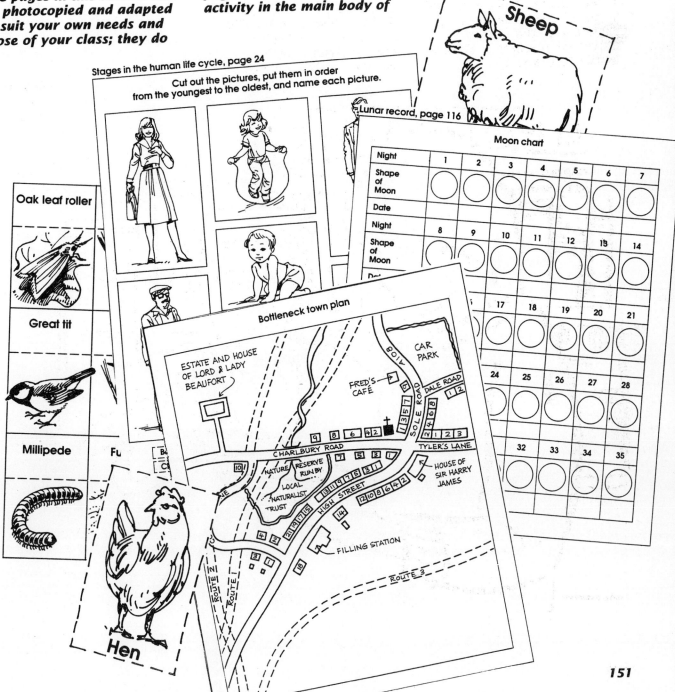

Sheep

Stages in the human life cycle, page 24

Cut out the pictures, put them in order from the youngest to the oldest, and name each picture.

Oak leaf roller

Great tit

Millipede

Lunar record, page 116

Moon chart

Night	1	2	3	4	5	6	7
Shape of Moon							
Date							
Night	8	9	10	11	12	13	14
Shape of Moon							
		17	18	19	20	21	
		24	25	26	27	28	
		32	33	34	35		

Bottleneck town plan

ESTATE AND HOUSE OF LORD & LADY BEAUFORT

FRED'S CAFÉ

CAR PARK

A108

DALE ROAD

SOLE ROAD

CHARLBURY ROAD

NATURE RESERVE RUN BY LOCAL NATURALIST TRUST

TYLER'S LANE

HOUSE OF SIR HARRY JAMES

HIGH STREET

FILLING STATION

ROUTE 2

ROUTE 1

ROUTE 3

Hen

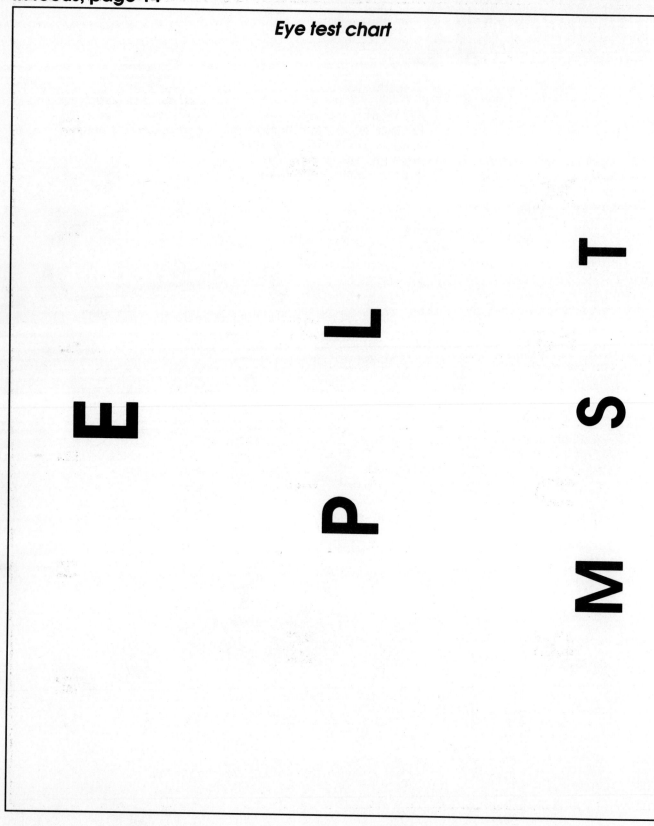

Eye test chart

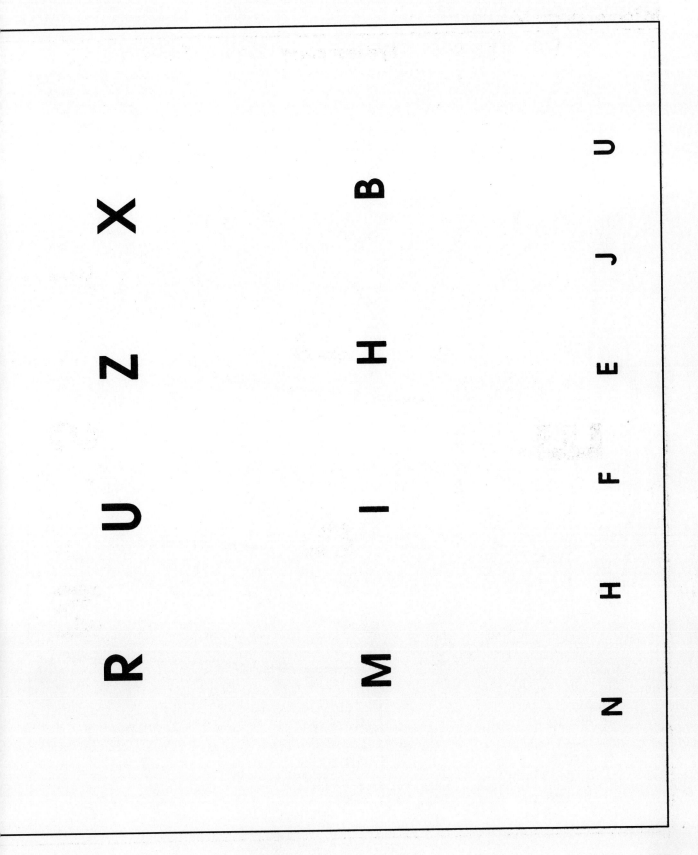

Cut out the parts of the body, put together the body and name each part

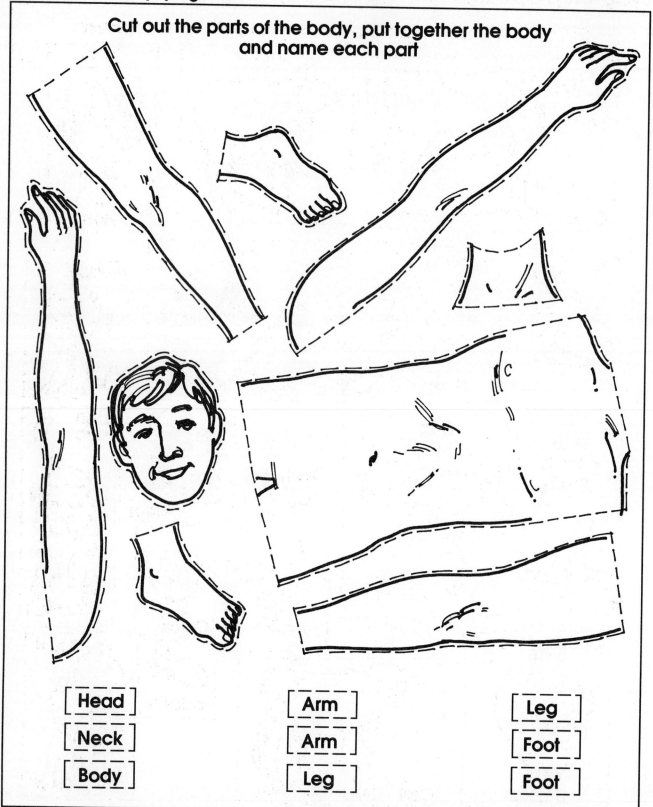

[Head]	[Arm]	[Leg]
[Neck]	[Arm]	[Foot]
[Body]	[Leg]	[Foot]

Cut out the animals and put the correct baby with its parent.

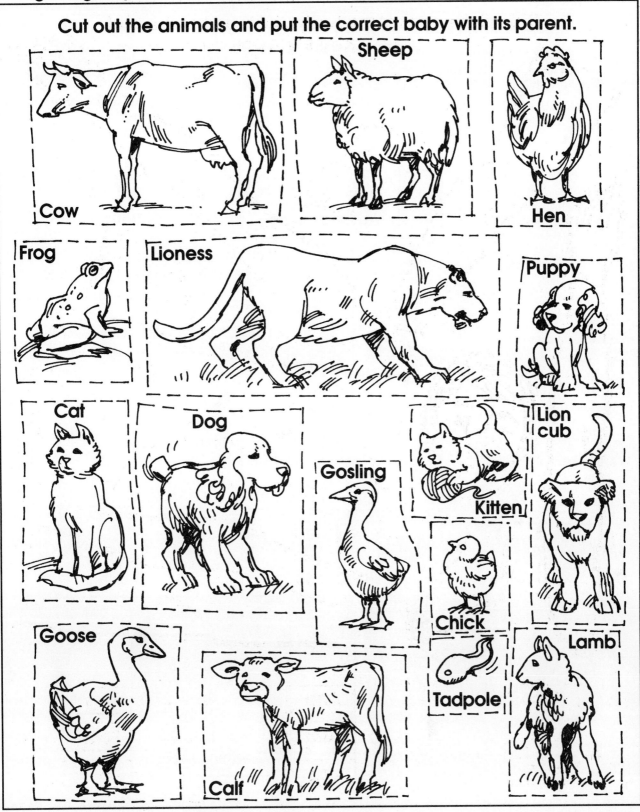

Cow

Sheep

Hen

Frog

Lioness

Puppy

Cat

Dog

Gosling

Kitten

Lion cub

Chick

Goose

Calf

Tadpole

Lamb

Cut out the pictures and put them in the correct order.
What did you do today?

Sleeping

Walking the dog

Playing

Eating

Watching television

Washing

Going to school

Drinking

Getting dressed

Cleaning teeth

Cut out the pictures, put them in order
from the youngest to the oldest, and name each picture.

[Baby]

[Child]

[Teenager]

[Mother]

[Father]

[Grandmother]

[Grandfather]

Cut out the outline of the body. Cut out the organs opposite and stick them on the body as numbered, ensuring that tabs are used when available.

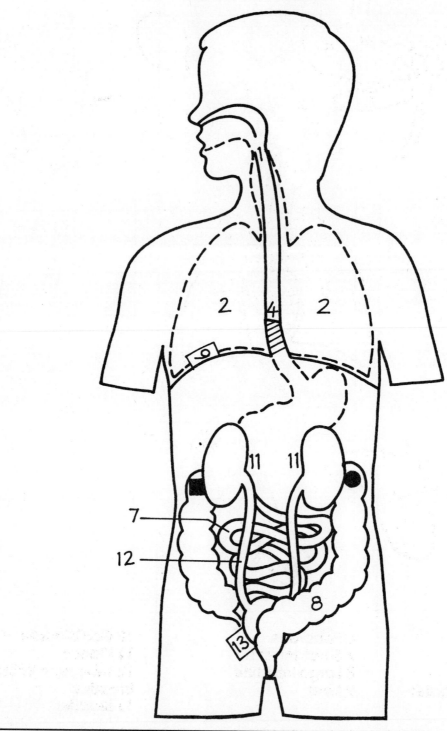

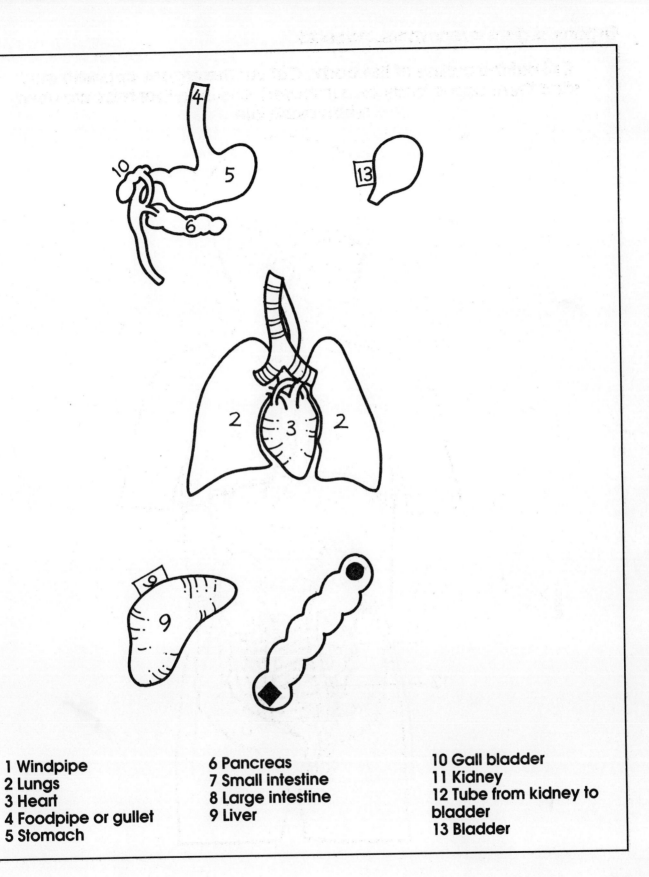

1 Windpipe
2 Lungs
3 Heart
4 Foodpipe or gullet
5 Stomach

6 Pancreas
7 Small intestine
8 Large intestine
9 Liver

10 Gall bladder
11 Kidney
12 Tube from kidney to bladder
13 Bladder

Cut out the diagram of the digestive system. Cut out the five statements opposite and use them to label the diagram by sticking them on as appropriate.

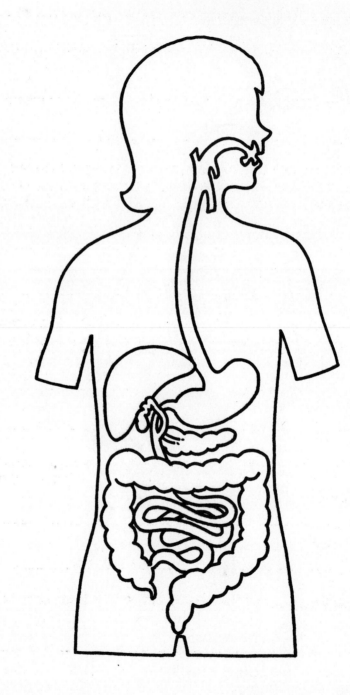

In the **stomach**, food is churned and more juices including acid are added.

In the **small intestine** more chemicals (called enzymes) are added. Proteins, carbohydrates and fats are finally broken down into small, soluble substances which pass through the wall of the small intestine and into the blood.

Waste food passes out of the body through the **anus**. This waste food is brown in colour and contains some moisture, but much of the goodness has been taken out of it.

In the **mouth**, the food is bitten, chewed and broken up into small pieces. Saliva is mixed with the food and the chemicals in it begin to break it down further.

In the **large intestine** much of the water in the food is taken into the blood.

Push or pull?

‑ ‑

‑ ‑

What can forces do?

Moves faster or slower	Changes shape or size	Changes direction

Sink or swim?

Floats	Sinks

Bridge testing sheet

Before testing

Estimate what load your bridge will hold.

_____ **g**

Where will the failure occur?

_____

Why there?

_____

After testing

What load did it hold?

_____ **g**

Where did it fail?

_____

Why did it fail here - if not as above?

_____

How could you change your design so that your
bridge would be stronger?

_____

Metals and plastics

Item for investigation / Test					
What colour is it?					
Is it shiny or dull?					
Is it smooth or rough?					
Does it feel cold or warm?					
Does it bend without breaking?					
Can you tear or cut it?					
Does it float or sink?					
Is it picked up by a magnet?					
Can it be scratched with a nail?					

Passage of electrical current in a circuit

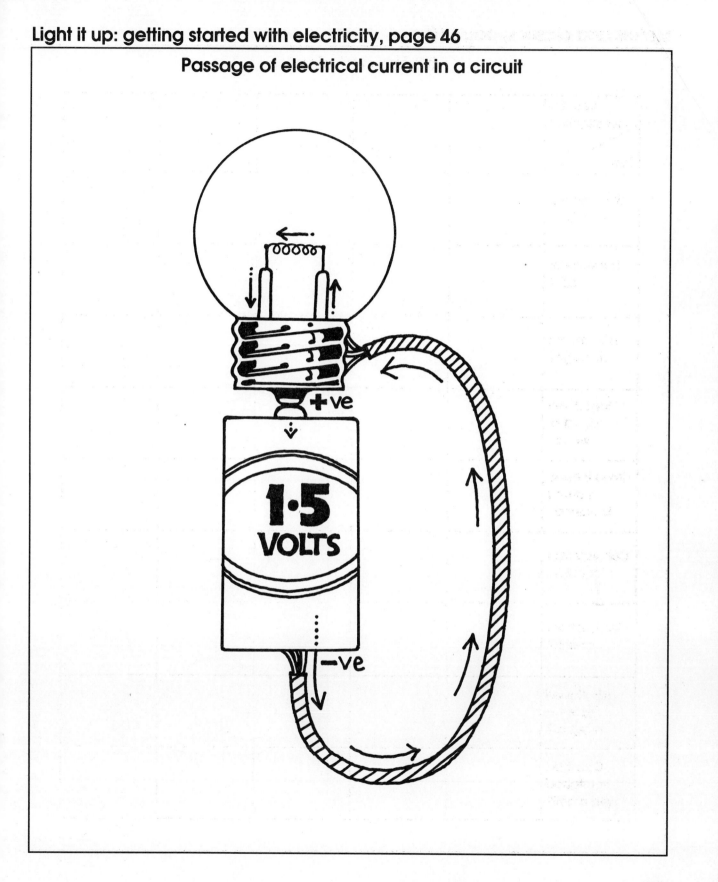

Will the bulb (lamp) light? That is the question!

A. Predict B. Record C. Test D. Record

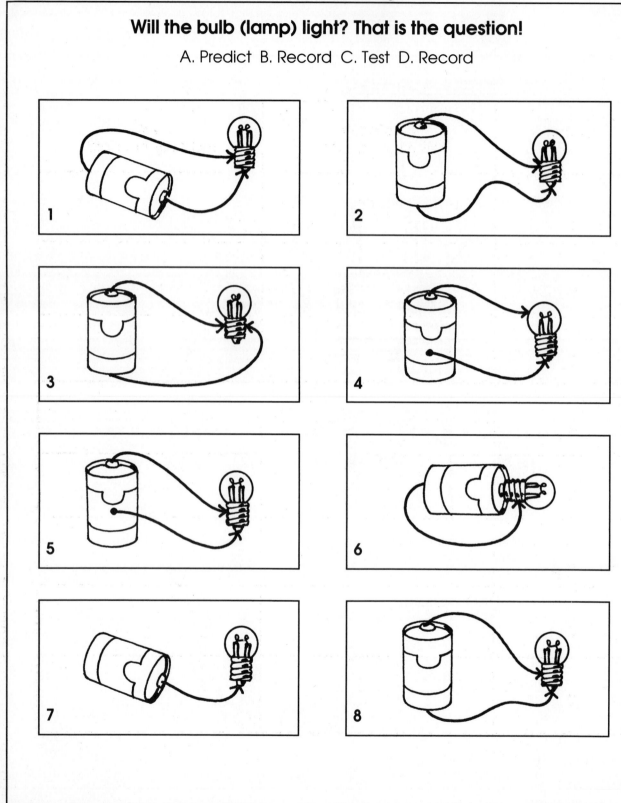

Circuit	Prediction Will it light or not?	Test Did it light or not?
1		
2		
3		
4		
5		
6		
7		
8		

Cut-out circuit pictures

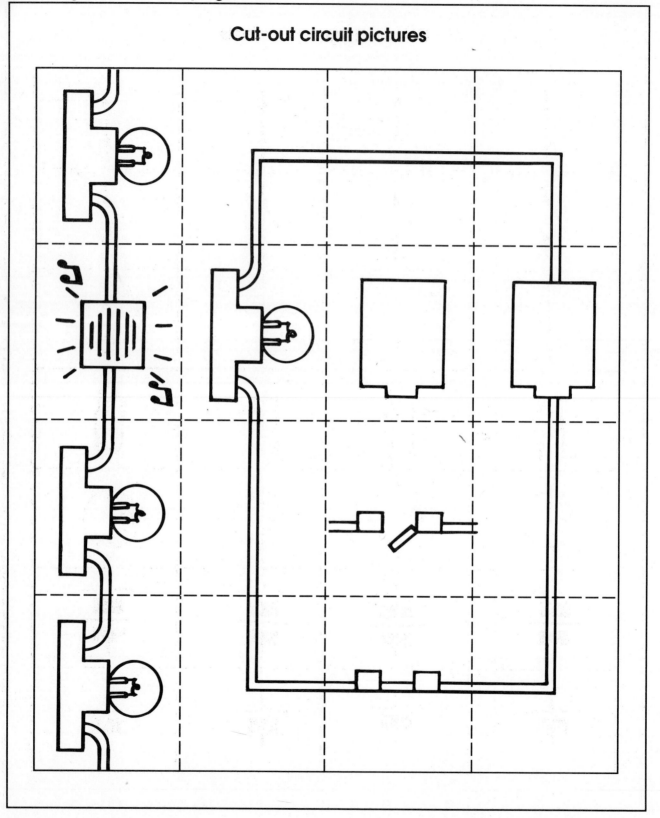

Circuit diagram symbols

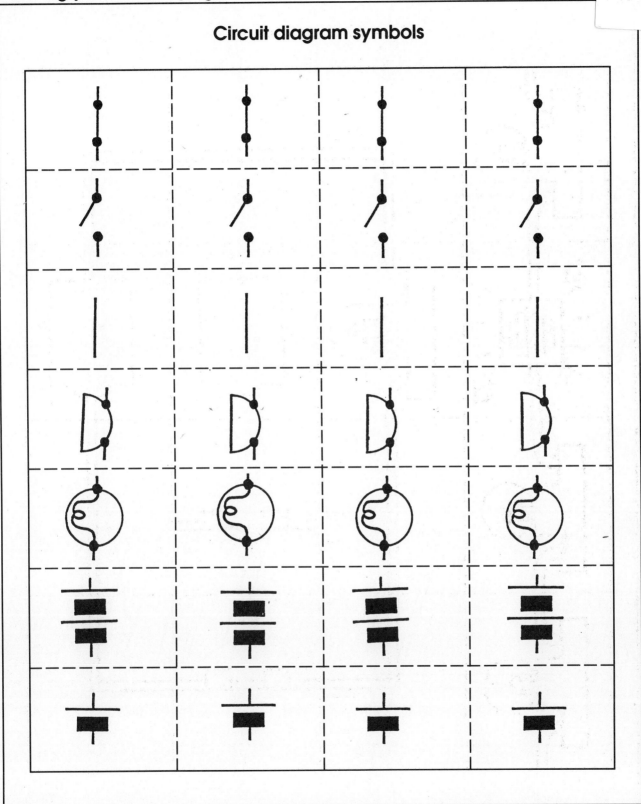

Making a moving structure from softwood and card

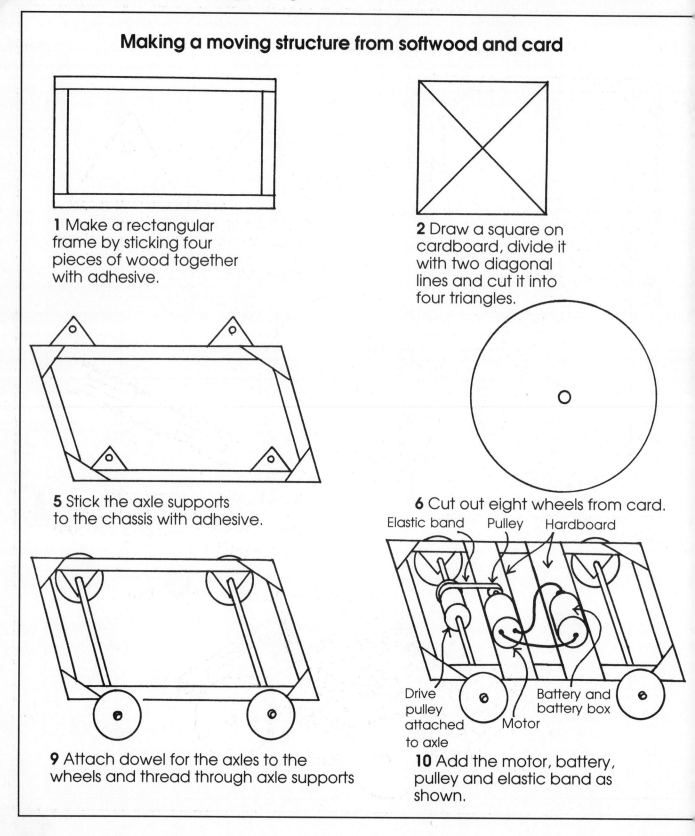

1 Make a rectangular frame by sticking four pieces of wood together with adhesive.

2 Draw a square on cardboard, divide it with two diagonal lines and cut it into four triangles.

5 Stick the axle supports to the chassis with adhesive.

6 Cut out eight wheels from card.

Elastic band   Pulley   Hardboard

Drive pulley attached to axle

Motor

Battery and battery box

9 Attach dowel for the axles to the wheels and thread through axle supports

10 Add the motor, battery, pulley and elastic band as shown.

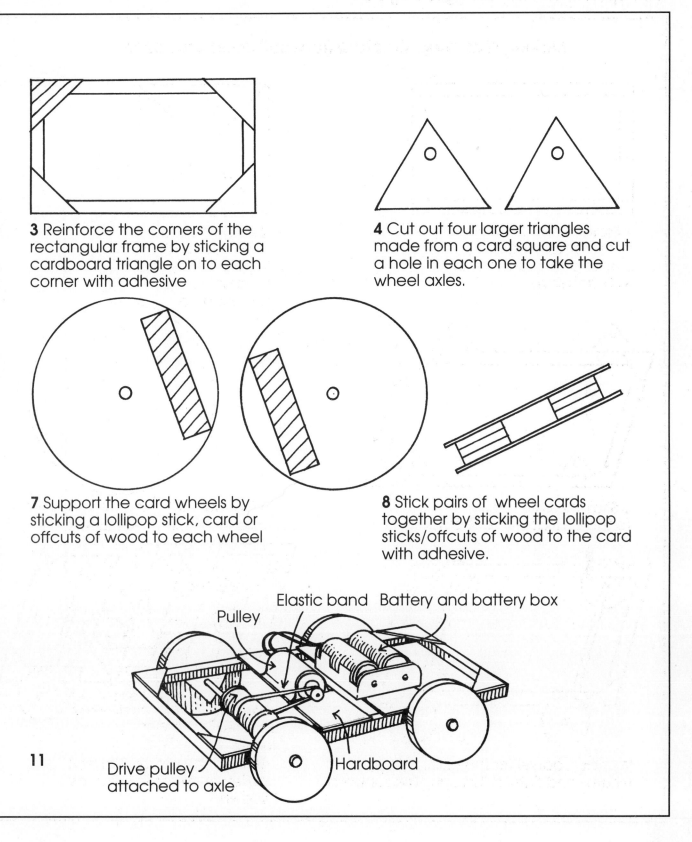

3 Reinforce the corners of the rectangular frame by sticking a cardboard triangle on to each corner with adhesive

4 Cut out four larger triangles made from a card square and cut a hole in each one to take the wheel axles.

7 Support the card wheels by sticking a lollipop stick, card or offcuts of wood to each wheel

8 Stick pairs of wheel cards together by sticking the lollipop sticks/offcuts of wood to the card with adhesive.

Elastic band   Battery and battery box

Pulley

11

Drive pulley
attached to axle

Hardboard

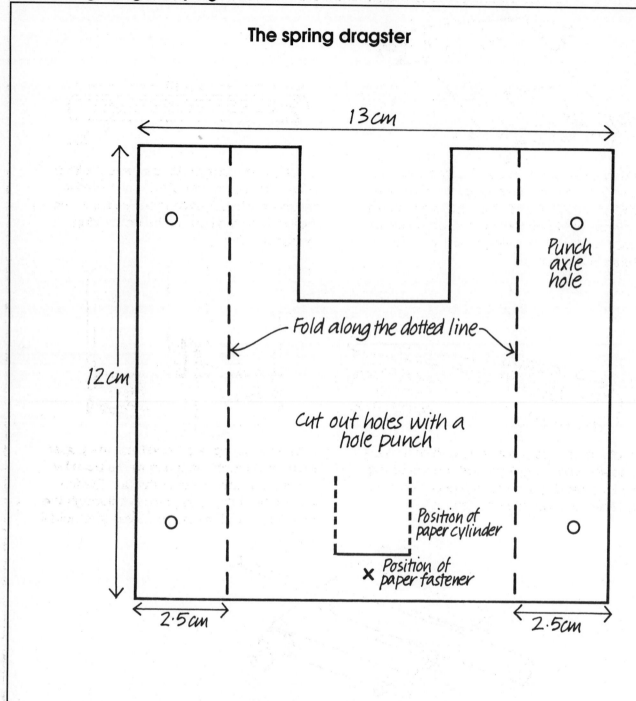

The spring dragster

13cm

Fold along the dotted line

Punch axle hole

12cm

Cut out holes with a hole punch

Position of paper cylinder

x Position of paper fastener

2·5cm   2·5cm

1 Cut out the outline of the dragster shown above, and stick it on to a piece of 1mm thick card. Accurately punch four holes where marked on the base using a hole punch and fold the sides of the base along the dotted lines.

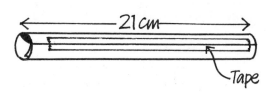

2 Using three cotton reels and two pieces of dowel (12cm and 18cm long) assemble the base as shown above. Put rubber grommets or pieces of plastic tubing or rolled up elastic bands on the ends of each axel to keep the wheels in position.

3 Role up a piece of A5 paper to make a tube with a diameter of 2cm and hold it together with adhesive tape. Stick it to the base where marked on the card, using adhesive tape.

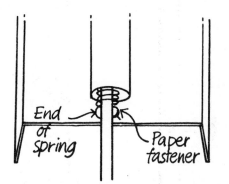

4 Make a hole 2cm from one end of a piece of dowel (18cm long) and attach the spring to it by threading one end through the hole and flattening out the end few coils.

5 Push the spring and dowel into the paper tube and attach the spring to the base by placing a paper fastener through the free end of the spring and pushing it through the card. Open out the paper fastener to hold it in place.

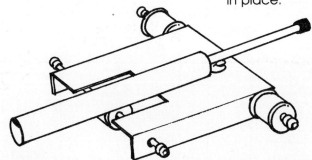

6 Put a rubber bung or piece of cork on to the end of the piece of dowel inside the cylinder. Hold the dragster so that the bung is pointing upwards and mark 0cm on the dowel where it touches the end of the cylinder. Use a ruler to mark 2cm, 4cm, 6cm and 8cm from the first line, moving towards the bung.

Weathering and erosion

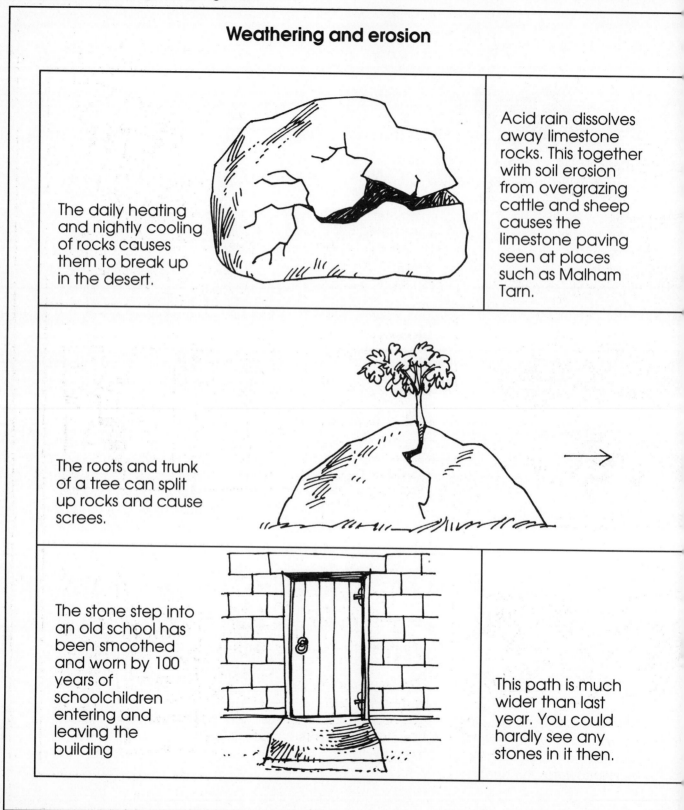

The daily heating and nightly cooling of rocks causes them to break up in the desert.

Acid rain dissolves away limestone rocks. This together with soil erosion from overgrazing cattle and sheep causes the limestone paving seen at places such as Malham Tarn.

The roots and trunk of a tree can split up rocks and cause screes.

The stone step into an old school has been smoothed and worn by 100 years of schoolchildren entering and leaving the building

This path is much wider than last year. You could hardly see any stones in it then.

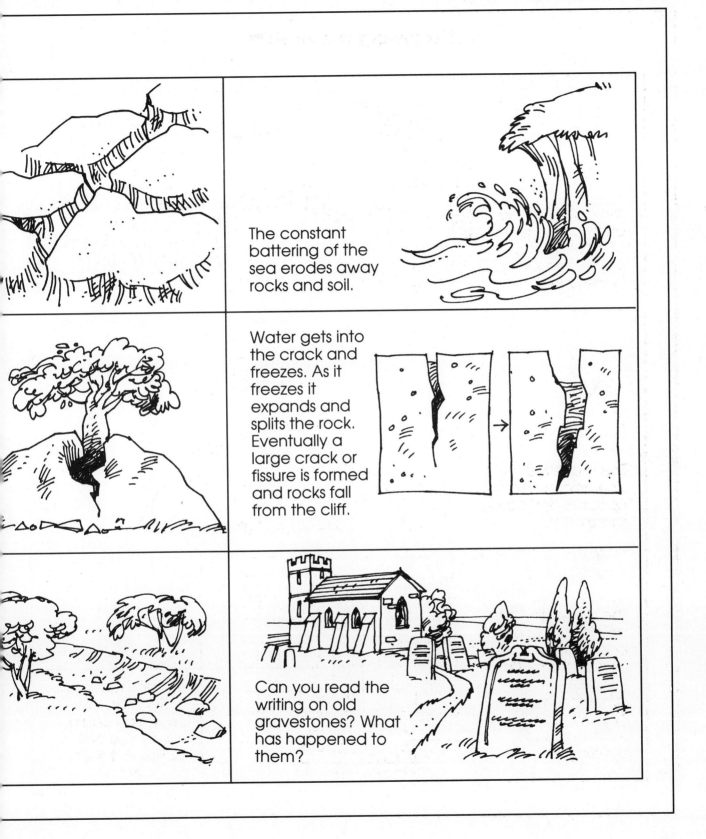

The constant battering of the sea erodes away rocks and soil.

Water gets into the crack and freezes. As it freezes it expands and splits the rock. Eventually a large crack or fissure is formed and rocks fall from the cliff.

Can you read the writing on old gravestones? What has happened to them?

Weather symbols

15 **2**	**Temperature**: Figures on a white background give positive temperature in degrees centigrade. Black figures on a shaded background give freezing temperatures, ie below zero centigrade.
25	**Sunshine:** A yellow symbol represents the sun; red figures in the centre show a temperature of 25 degrees centigrade.
	Cloud: A white cloud symbol indicates fine weather clouds that may be relatively thin and patchy.
	Cloud: A black cloud represents the thicker and more widespread clouds often associated with dull weather.
	Sunny intervals: The sun symbol used in conjunction with a cloud in this way means some sunshine as well as cloud, particularly if the white cloud symbol is used.
	Rain: The tear drop symbols beneath the cloud indicate rain.

	Rain showers and sunny intervals: A combination of rain, cloud and sun represents sunny intervals and rain showers.
	Snow: The snow symbols beneath the cloud indicate snow.
	Sleet: The rain and snow symbols together beneath the cloud indicate sleet.
	Thunder storm: The symbol of a cloud with a flash represents the possibility of thunder and lightning.
	Wind speed and direction: The black symbol represents the wind speed and direction. The speed printed in the centre in white is in miles per hour.
FOG	**Fog:** Fog is not represented by a specific symbol, it is indicated by words on the map in the general areas likely to be affected.
MIST	**Mist:** Mist is not represented by a specific symbol, it is indicated by words on the map in the general areas likely to be affected.

Weather charts

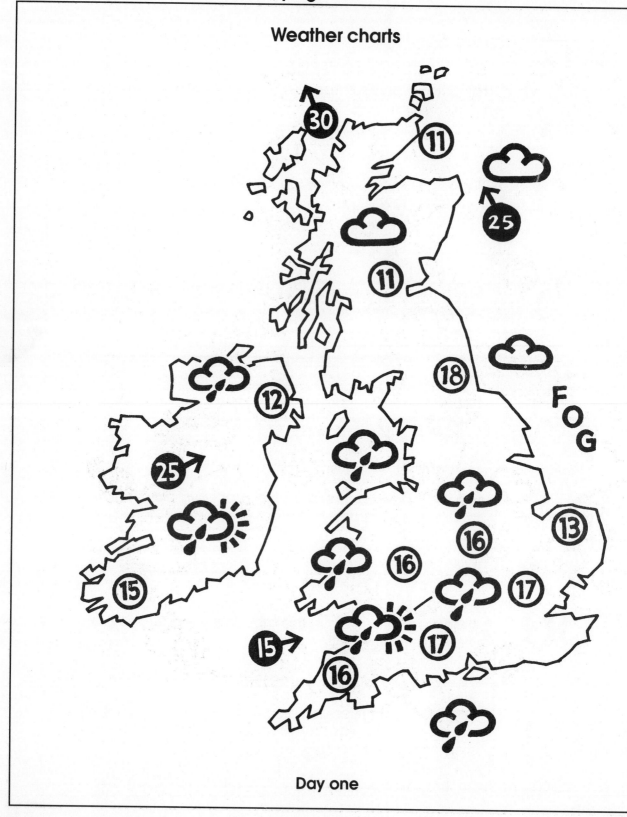

Day one

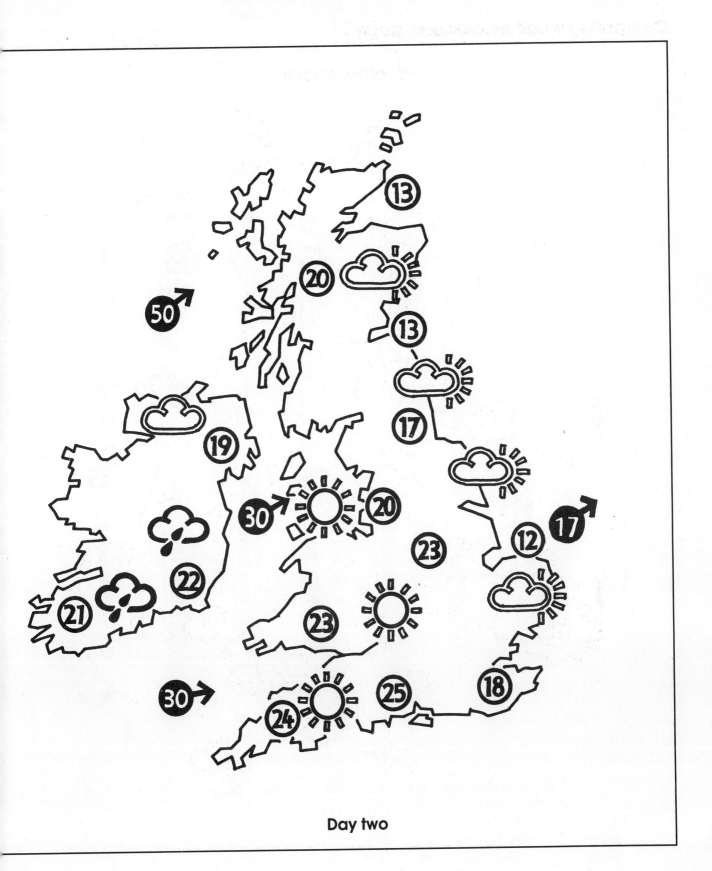

Day two

Measuring temperature

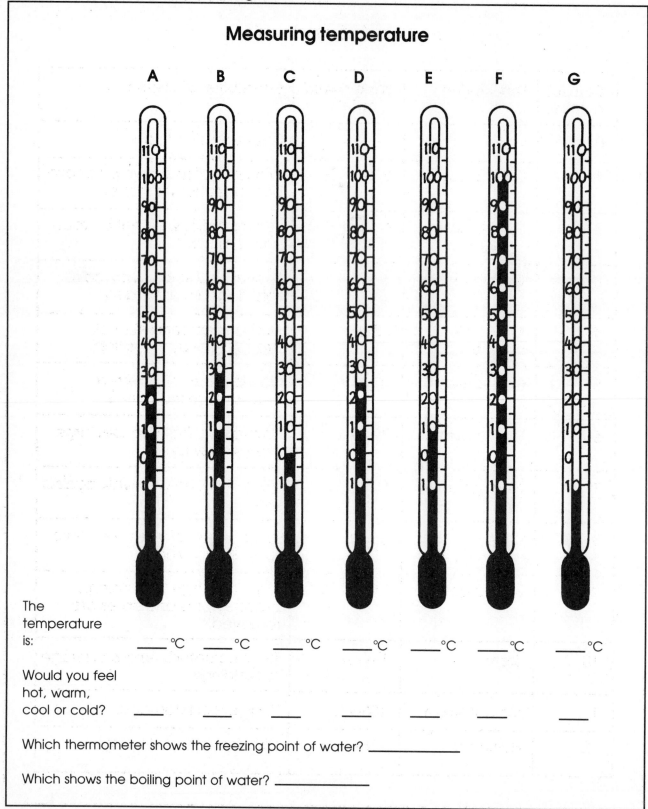

The temperature is:

____°C   ____°C   ____°C   ____°C   ____°C   ____°C   ____°C

Would you feel hot, warm, cool or cold?

____   ____   ____   ____   ____   ____   ____

Which thermometer shows the freezing point of water? _____

Which shows the boiling point of water? _____

Beaufort wind scale

Beaufort number	Description	Wind speed (km/hr)	Description of effects
0	Calm	0-1	Smoke rises vertically.
1	Light air	2-5	Wind direction shown by smoke, but vanes do not move.
2	Light breeze	6-11	Leaves rustle, wind felt on face, vanes move.
3	Gentle breeze	12-19	Leaves and small twigs move, light flags are extended.
4	Moderate breeze	20-28	Dust and paper rise, small branches sway, flags flap.
5	Fresh breeze	29-38	Small trees sway, crested waves on inland waters.
6	Strong breeze	39-49	Umbrellas difficult to use, large branches sway.
7	Near gale	50-61	Trees sway, hard to walk against wind.
8	Gale	62-74	Twigs break off trees, very hard to walk into wind.
9	Strong gale	75-88	Slight damage to buildings, chimney pots and slates are removed.
10	Storm	89-102	Trees uprooted, serious damage to buildings.
11	Violent storm	103-117	Widespread damage.
12	Hurricane	118+	Disaster, terrible damage.

Bottleneck town plan

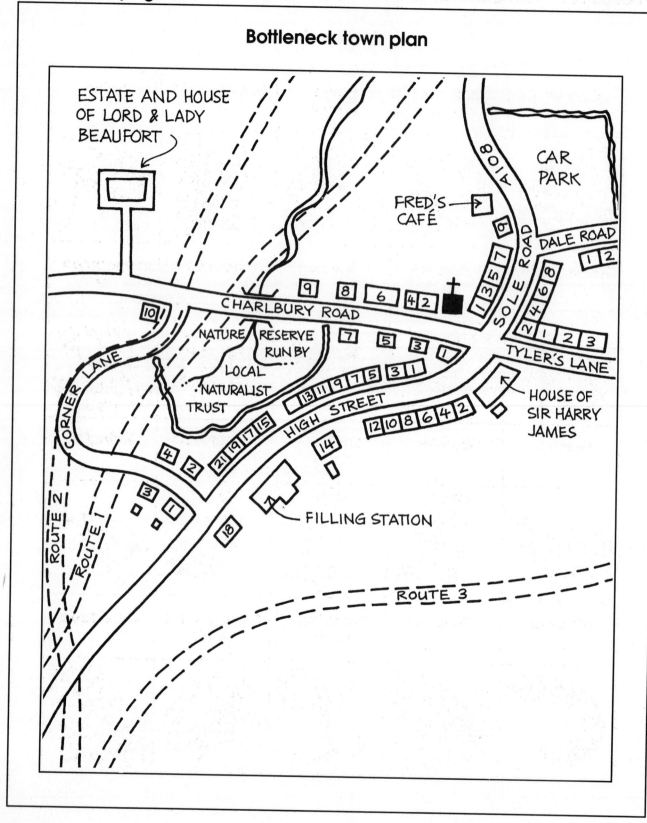

ESTATE AND HOUSE OF LORD & LADY BEAUFORT

CAR PARK

FRED'S CAFÉ

CHARLBURY ROAD

SOLE ROAD

A108

DALE ROAD

TYLER'S LANE

HOUSE OF SIR HARRY JAMES

NATURE RESERVE RUN BY LOCAL NATURALIST TRUST

CORNER LANE

HIGH STREET

FILLING STATION

ROUTE 2

ROUTE 1

ROUTE 3

Food web

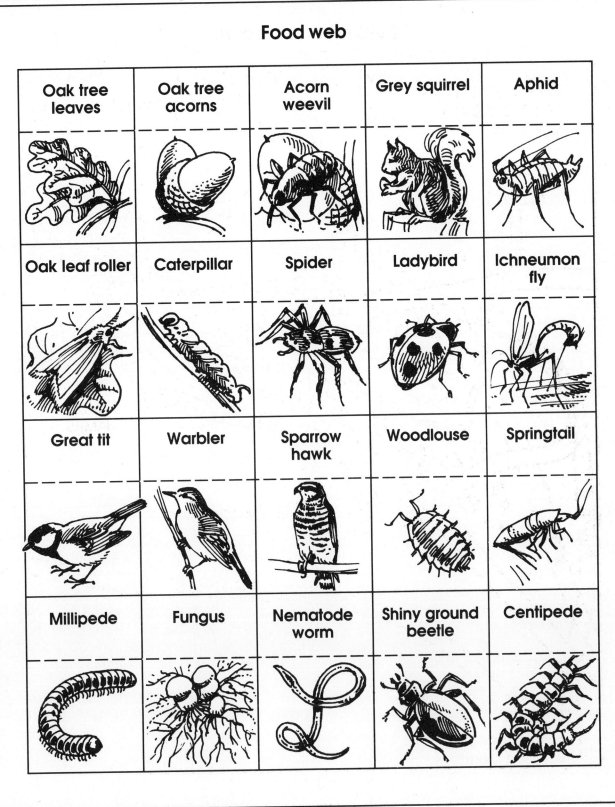

Oak tree leaves	Oak tree acorns	Acorn weevil	Grey squirrel	Aphid
Oak leaf roller	Caterpillar	Spider	Ladybird	Ichneumon fly
Great tit	Warbler	Sparrow hawk	Woodlouse	Springtail
Millipede	Fungus	Nematode worm	Shiny ground beetle	Centipede

Clouds produce rainfall.

The rain finds its way into streams and rivers.

Some water is used by man but most returns to the rivers or evaporates.

Evaporation from the river.

Evaporation from the sea.

Rivers flow into the sea.

The heat from the sun causes water in the sea and other areas to evaporate as water vapour.

As water vapour rises it is cooled. The water vapour condenses into tiny drops of water whichform clouds.

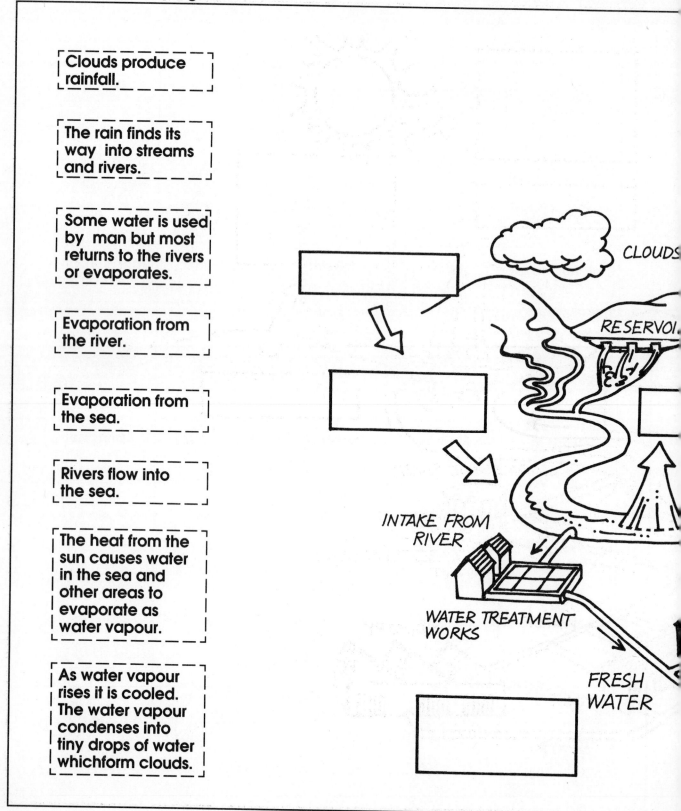

CLOUDS

RESERVOI.

INTAKE FROM RIVER

WATER TREATMENT WORKS

FRESH WATER

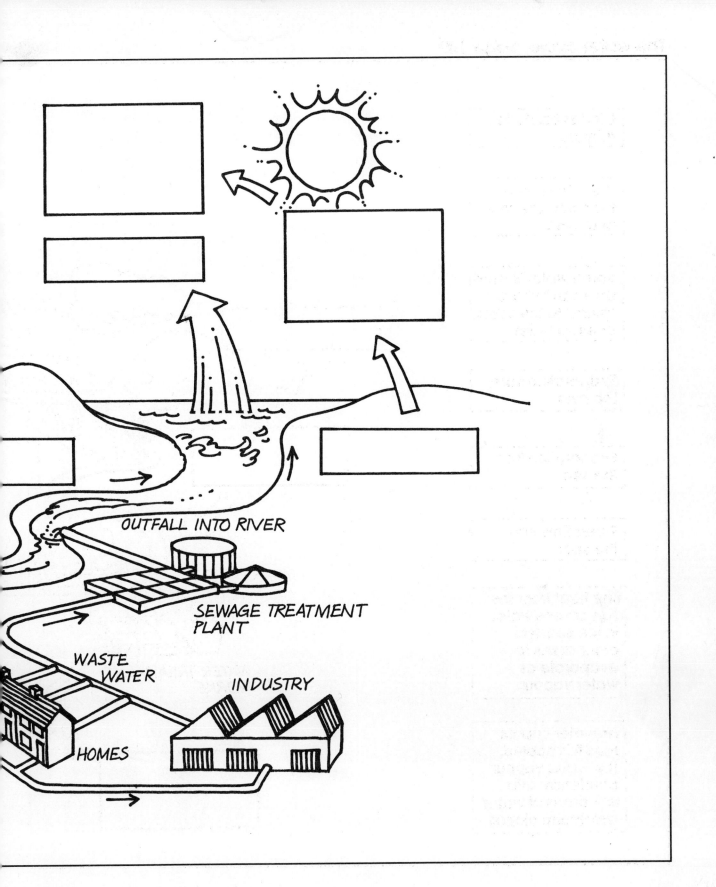

OUTFALL INTO RIVER

SEWAGE TREATMENT
PLANT

WASTE
WATER

INDUSTRY

HOMES

Four seasons

Moon chart

Night	1	2	3	4	5	6	7
Shape of Moon	○	○	○	○	○	○	○
Date							
Night	8	9	10	11	12	13	14
Shape of Moon	○	○	○	○	○	○	○
Date							
Night	15	16	17	18	19	20	21
Shape of Moon	○	○	○	○	○	○	○
Date							
Night	22	23	24	25	26	27	28
Shape of Moon	○	○	○	○	○	○	○
Date							
Night	29	30	31	32	33	34	35
Shape of Moon	○	○	○	○	○	○	○
Date							